EMAS 2023 Workshop - 17th European Workshop on Modern Developments and Applications in Microbeam Analysis

IOP Conference Series: Materials Science and Engineering
Volume 1324

Krakow, Poland
7-11 May 2023

ISBN: 979-8-3313-1417-0
ISSN: 1757-8981

Printed from e-media with permission by:

Curran Associates, Inc.
57 Morehouse Lane
Red Hook, NY 12571

Institute of Physics
Dirac House, Temple Back
Bristol BS1 6BE UK

Phone: 44 1 17 929 7481
Fax: 44 1 17 920 0979

techtracking@iop.org

Additional copies of this publication are available from:

Curran Associates, Inc.
57 Morehouse Lane
Red Hook, NY 12571 USA
Phone: 845-758-0400
Fax: 845-758-2633
Email: curran@proceedings.com
Web: www.proceedings.com

TABLE OF CONTENTS

17th European Workshop on Modern Developments and Applications in Microbeam Analysis (EMAS 2023) Krakow, Poland, 7-11 May 2023

Marek Faryna[1], Grzegorz Tylko[2], Bjørn E. Sørensen[3], Hans Dijkstra[4] and Silvia Richter[4]

[1] Polish Academy of Sciences, Alek. Krupkowski Inst. Metallurgy & Materials Science, ul. W. Reymonta 25, 30059 Krakow, Poland

[2] Jagiellonian University, Inst. Zoology & Biomed. Research, Dept. Cell Biology & Imaging, ul. Gronostajowa 9, 30387 Krakow, Poland

[3] Norwegian University of Science and Technology, Department of Geoscience and Petroleum, 7491 Trondheim, Norway

[4] Thermo Fisher Scientific BV, Breda, The Netherlands

[5] RWTH Aachen, Gemeinschaftslabor für Elektronenmikroskopie (GfE), 552074 Aachen, Germany

Preface

This volume of the IOP Conference Series: Materials Science and Engineering contains papers from the 17th Workshop of the European Microbeam Analysis Society (EMAS) on Modern Developments and Applications in Microbeam Analysis, which took place from the 7th to the 11th of May 2023 at the Auditorium Maximum of the Jagiellonian University, Krakow, Poland.

The primary aim of this series of workshops is to assess the state-of-the-art and reliability of microbeam analysis techniques. The workshops also provide a forum where students and young scientists starting out on a career in microbeam analysis can meet and discuss with the established experts. The Krakow meeting was organised within the usual EMAS format, comprising invited plenary lectures by internationally recognised experts, poster presentations by the participants and round-table discussions on the key topics led by specialists in the field. EMAS2023 was organized in collaboration with the Institute of Metallurgy and Materials Science of the Polish Academy of Sciences, Krakow, Poland. The technical programme included the following topics: Electron probe microanalysis (EPMA); Electron backscatter diffraction (EBSD); Software tools; Focussed ion beam; Combined techniques in SEM; and materials applications of microbeam analysis and their applications.

As at previous workshops, there was also a special oral session for Early Career Scientists. The best presentation by an early career scientist was awarded with an invitation to attend the 2024 Microscopy and Microanalysis meeting in Cleveland, Ohio. The prize went to Maria Wątroba, EMPA - Swiss Federal Labs. Materials Testing and Research, Thun (CH), for her presentation entitled: "*Experimental analysis of plastic deformation in single-crystalline and ultrafine-grained Zn micropillars*" (by M. Wątroba, W. Bednarczyk, C. Tian, J. Michler and J. Schwiedrzik).

IOP Conf. Series: Materials Science and Engineering **1324** (2025) 011001 doi:10.1088/1757-899X/1324/1/011001

Three half- or full-day pre-meeting courses were held covering the following areas of microbeam analysis: Introduction to EBSD, Introduction to X-ray microanalysis, Monte Carlo simulation, and three post-meeting working group meetings were devoted to EPMA of Earth Science materials, Facility management, and EPMA optimisation in irradiated and nuclear material analysis.

The continuing relevance of the EMAS meetings and the high regard in which they are held internationally can be seen from the fact that 70 posters from 13 countries were on display at the meeting and that the participants came from as far away as Japan and U.S.A. A selection of participants with posters was invited to give a short oral presentation of their work in three dedicated sessions. The EMAS Early Career Scientist Poster Award was awarded with an invitation to participate in the AMAS XVI: the Australian Microanalysis Society's biennial symposium, February 5-9, 2024, at the Queensland University of Technology (QUT), in Brisbane, Australia. The prize was awarded to Amelia Zięba, AGH - University of Science and Technology, Academic Centre for Materials and Nanotechnology, Krakow (PL), for her contribution entitled: "*Characterisation of ternary intermetallic alloys for catalytic applications*" (by A. Zięba, K. Stan-Głowińska, Ł. Rogal, D. Duraczyńska and L. Lityńska Dobrzyńska). The EMAS Workshop Poster Prize was given to Tobias Salge, Natural History Museum, London (UK), for his contribution entitled: "*Non-destructive, non-invasive SEM/EDS of the Winchcombe meteorite using an annular SDD*" (by T. Salge, A.J. King and S.S. Russell).

This proceedings volume contains the full texts of 8 of the invited plenary lectures and of 7 papers on related topics originating from the posters presented at the workshop. All the papers have been subjected to peer review by a least two referees.

September 2024

EMAS 2023 Workshop, held at the Auditorium Maximum of the Jagiellonian University, Krakow, Poland, from 7 to 11 May 2023.

Acknowledgements
On behalf of the European Microbeam Analysis Society I would like to thank all the invited speakers, session chairs and members of the discussion panels for making the meeting such a great success. Special thanks go to Marek Faryna, Grzegorz Tylko and Luc Van 't dack who directed the organisation of the workshop giving freely of their time and talents. As was the case for previous workshops, the EMAS Board *in corpore* was responsible for the scientific programme. The Workshop also included a commercial exhibition where many leading instrument suppliers were represented. Several companies that exhibited provided financial support, either by sponsoring an event or by advertising.

Below, in alphabetical order, is a list of exhibiting companies and sponsors of the workshop:

- Ametek GmbH, GATAN-EDAX Business Unit
- Ametek, CAMECA Business Unit
- Carl Zeiss Sp.z.o.o.
- Delmic BV
- Guillaume F. Bastin Fund of the European Microbeam Analysis Society (EMAS) sponsored by: Thermo Fisher Scientific
- Jagiellonian University
- JEOL (Europe) SAS
- Micro to Nano
- NewTec Scientific
- Oxford Instruments GmbH
- Polish Academy of Sciences, Institute of Metallurgy and Materials Science
- Target-Messtechnik
- Technolutions Sp.z.o.o.
- TESCAN ORSAY HOLDING, a.s.
- Thermo Fisher Scientific

Hans Dijkstra
EMAS President

IOP Conf. Series: Materials Science and Engineering **1324** (2025) 011002 doi:10.1088/1757-899X/1324/1/011002

Peer Review Statement

All papers published in this volume have been reviewed through processes administered by the Editors. Reviews were conducted by expert referees to the professional and scientific standards expected of a proceedings journal published by IOP Publishing.

- **Type of peer review:** Double Anonymous
- **Conference submission management system:** Morressier
- **Number of submissions received:** 15
- **Number of submissions sent for review:** 15
- **Number of submissions accepted:** 15
- **Acceptance Rate (Submissions Accepted / Submissions Received × 100):** 100
- **Average number of reviews per paper:** 2
- **Total number of reviewers involved:** 20
- **Contact person for queries:**
  Name: Luc Van't dack
  Email: luc.vantdack@uantwerpen.be
  Affiliation: EMAS - European Microbeam Analysis Society

A new empirical model of the 2-dimensional ionisation probability $\phi(\rho r, \rho z)$ described by an analytical function

G Achuda[1], J Bünger[2], P T Pinard[3], M Torrilhon[2] and S Richter[1]

[1] RWTH Aachen University, Central Facility for Electron Microscopy, Aachen, Germany
[2] RWTH Aachen University, Applied and Computational Mathematics, Aachen, Germany
[3] Oxford Instruments NanoAnalysis, High Wycombe, United Kingdom

E-mail: achuda@gfe.rwth-aachen.de

Abstract. For quantitative electron probe microanalysis matrix effects have to be considered. Conventional matrix correction procedures in use today assume a homogeneous composition for the bulk sample (known as ZAF correction) or heterogeneity only in depth for thin films ($\phi(\rho z)$ models). But when heterogeneity is expected in lateral directions, only Monte Carlo and deterministic simulations could be used for accurate matrix correction. These simulations are computationally expensive and slow. In this work, we introduce a novel analytical model for the multi-dimensional ionisation probability distribution $\phi(\rho r, \rho z)$ for a material bombarded with electrons, which could be used for fast matrix correction as an alternative to simulations. By fitting the model onto ionisation probability distributions generated by Monte Carlo simulations, we show that our model can adequately describe the simulated distributions.

1. Introduction

Electron probe X-ray microanalysis is a well-established method for the chemical analysis of materials on the micron to sub-micrometre scale. The characterisation of advanced materials and geological samples often requires accurate determination of the chemical composition of small phases (e.g., particles, precipitates, inclusions) and elemental segregation near phase boundaries. While electron microscopes equipped with field emission guns can deliver nanometre-sized beam diameters, the size of the X-ray emission volume could be larger than the size of the features to be analysed. Further, for accurate quantification of materials, matrix affects such as material-dependent scattering of electrons, absorption of X-rays and secondary X-ray fluorescence need to be corrected for. The two conventional matrix correction procedures, ZAF and $\phi(\rho z)$-based correction models, are, strictly only applicable when the sample is homogeneous within the X-ray emission volume. Some $\phi(\rho z)$-based models have been extended for the analysis of multilayered samples [1-3] using weighting procedures. Samples with heterogeneities in their lateral dimensions require reconstruction techniques which take into account the lateral spread of the X-ray generation volume. Inverse modelling of Monte Carlo simulations [4-6] can be an option, but the presence of stochastic noise makes them poorly suited for gradient based optimisation methods. Recently developed fast deterministic models [7-9] have no stochastic noise, but they are still not computationally light. Faster compute times would be beneficial when a large number of measurements, for example from line-scans and maps, need to be inverted, or to find good initial guesses for the more compute-intensive methods. Hence, the goal of this work is to develop an analytical model to describe the three-dimensional (3D) ionisation probability distribution in a material subject to an electron beam, an extension of the famous $\phi(\rho z)$-function.

2. The multi-dimensional ionisation probability distribution

2.1. Theoretical considerations

The one-dimensional (1D) $\phi(\rho z)$ is defined as the ratio of intensity of X-rays generated from a thin horizontal slice of the material at depth ρz to that of the same slice if it was unsupported [10]. The number of primary ionisations of a sub-shell l of element A, $\mathcal{N}_{A,l}^{ion}(\rho z)$, in a thin horizontal slice of the sample of mass-thickness $d\rho z$ at mass-depth ρz is related to the corresponding $\phi_{A,l}(\rho z)$ as

$$\mathcal{N}_{A,l}^{ion}(\rho z) = \mathcal{N}_{el}\left(\frac{N_0}{A_A}\right) Q_{A,l}(E_0)c_A(\rho z)\phi_{A,l}(\rho z)\mathrm{d}\rho z \tag{1}$$

where N_{el} represents the number of incident electrons, E_0 their initial energy, $c_A(\rho z)$ the mass fraction of element A at depth ρz, $Q_{A,l}(E)$ the cross-section for ionisation of the atomic sub-shell (A, l) by electrons of energy E, N_0 the Avagadro Constant, and A_A the atomic mass of A. Similarly, we can suppose that the 3D $\phi_{A,l}(\rho x, \rho y, \rho z)$ gives the number of primary ionisations of (A, l) in a small cube of dimensions $d\rho x \times d\rho y \times d\rho z$ at a position $(\rho x, \rho y, \rho z)$ as follows

$$\mathcal{N}_{A,l}^{ion}(\rho x, \rho y, \rho z) = \mathcal{N}_{el}\left(\frac{N_0}{A_A}\right) Q_{A,l}(E_0)c_A(\rho x, \rho y, \rho z)\phi_{A,l}(\rho x, \rho y, \rho z)\mathrm{d}\rho x\mathrm{d}\rho y\mathrm{d}\rho z. \tag{2}$$

For X-rays resulting from the ionisation of the sub-shell (A, l), the detected intensity, $I_{A,l}^{det}$, is then given by

$$I_{A,l}^{det} = C\mathcal{N}_{el}Q_{A,l}(E_0)\int\int\int c_A(\rho x, \rho y, \rho z)\phi_{A,l}(\rho x, \rho y, \rho z)\mathcal{A}(\rho x, \rho y, \rho z)\mathrm{d}\rho x\mathrm{d}\rho y\mathrm{d}\rho z \tag{3}$$

where, C is a constant that accounts for the fluorescence yield and the detector properties, and $\mathcal{A}(\rho x, \rho y, \rho z)$ accounts for the absorption of X-rays on their path to the detector. Absorption can be approximated using the Beer-Lambert law as

$$\mathcal{A}(\rho x, \rho y, \rho z) = \exp\left(-\int_{\mathcal{P}(\rho x, \rho y, \rho z)} \mu_m(\tilde{\rho x}, \tilde{\rho y}, \tilde{\rho z})\mathrm{d}\rho s\right) \tag{4}$$

where, $\mathcal{P}(\rho x, \rho y, \rho z)$ is the straight line connecting the point $(\rho x, \rho y, \rho z)$ to the detector and ρs represents the distance travelled along this line.

Analytical functions are only suitable for homogeneous samples because it would be prohibitively complex to consider the infinitely many ways in which a sample can be heterogeneous. Hence it is more convenient to extend the models meant for homogeneous samples to heterogeneous samples via appropriate weighting procedures. Such a weighting procedure would determine a fictitious homogeneous sample whose ionisation distribution is expected to closely resemble that of the heterogeneous sample of interest. This is done by averaging out the composition in the sample by weighting each point in the sample based on how likely it is to influence the final distribution.

However, for homogeneous samples, $\phi(\rho x, \rho y, \rho z)$ exhibits radial symmetry about the Z axis. Therefore, we can represent the distribution adequately using the two-dimensional (2D), $\phi(\rho r, \rho z)$, where ρr is the radial mass distance from the axis of the beam. $\phi(\rho r, \rho z)$ is related to $\phi(\rho x, \rho y, \rho z)$ as

$$\phi(\rho x, \rho y, \rho z) = \phi\left(\rho r = \sqrt{(\rho x)^2 + (\rho y)^2}, \rho z\right). \tag{5}$$

2.2. *Generating a $\phi(\rho r, \rho z)$ database*

In the past, tracer techniques have been used to experimentally determine $\phi(\rho z)$ distributions [11, 12]. Here a thin slice of material, called the tracer, containing a tracer element is embedded within a substrate which is devoid of the tracer element. $\phi(\rho z)$ is then calculated from the intensities of X-rays emitted by the tracer element which are measured as the tracer is slowly revealed by etching away layers of the substrate. It is not practical to use such experiments to generate a $\phi(\rho r, \rho z)$ or $\phi(\rho x, \rho y, \rho z)$ database, due to the complexity of preparing suitable tracers, the small emission volume of such tracers and the large number of measurements required to fill the multidimensional grids for these distributions. Hence, we chose to work with synthetic $\phi(\rho r, \rho z)$ data generated by Monte Carlo Simulations. In the Monte Carlo method, individual electron trajectories are simulated as random walks of electrons in the material with interactions modelled as random events based on their respective cross-sections. $\phi(\rho r, \rho z)$ can be computed from these simulated trajectories. The NISTMonte Monte-Carlo [13] package included in the NIST DTSA-II software was chosen as the Monte Carlo tool to obtain the $\phi(\rho r, \rho z)$ distributions, as the source code is publicly available and is easily modifiable.

To compute $\phi(\rho r, \rho z)$ from simulated electron trajectories, we first discretise the material into rings in ρr and ρz as shown in Figure 1. We also discretise energy in the range of E_0 to $E_{A,l}$, where E_0 is the electron beam energy and $E_{A,l}$ is the binding energy of the atomic sub-shell (A, l). We can now compute the electron density distribution as a 3 dimensional histogram, N, from the simulated electron trajectories as

$$N_{ijk} = \frac{\begin{array}{c}\text{the average distance travelled by an electron in the volume bin (i,j)} \\ \text{as shown in Figure 1b with energy between } E_k - \Delta E_k/2 \text{ and } E_k + \Delta E_k/2\end{array}}{\Delta\rho r_i \Delta\rho z_j \Delta E_k} \tag{6}$$

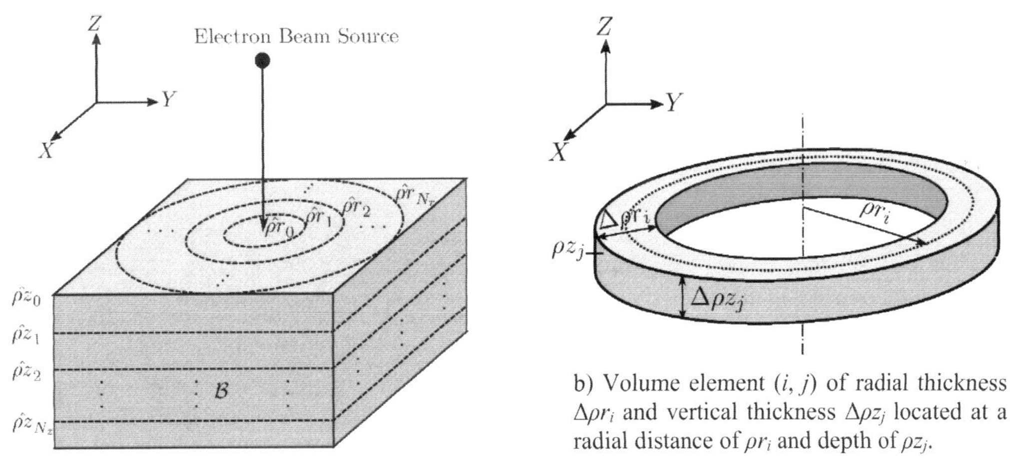

a) All bins.

b) Volume element (i, j) of radial thickness $\Delta\rho r_i$ and vertical thickness $\Delta\rho z_j$ located at a radial distance of ρr_i and depth of ρz_j.

Figure 1. a) Discretisation of the radial and vertical distances - ρr and ρz respectively. For any quantity, say $*$, $\hat{*}_{i-1}$ and $\hat{*}_i$ represent endpoints of the ith bin, $*_i$ its midpoint, and $\Delta *_i$ its width.

where N_{ijk} is the value at the index (i, j, k) in N. $\phi(\rho r, \rho z)$ can then be approximated as a histogram, whose $(i, j)^{\text{th}}$ component, corresponding to $\phi(\rho r_i, \rho z_j)$, is

$$\phi(\rho r_i, \rho z_j) \approx \phi_{ij} = \frac{1}{Q_{A,l}(E_0)} \sum_k Q_{A,l}(E_k) N_{ijk} \Delta E_k. \tag{7}$$

In this fashion, we generated data for the characteristic X-ray and over-voltage ratio (*U*) combinations tabulated in Table 1 in matrix elements listed in Table 2. Here, over-voltage ratio is the ratio of the beam energy, E_0, to the minimum energy required to ionise the atomic sub-shell which results in the characteristic X-ray of interest. This results in a total of (6×6−3)×24 = 792 unique experimental set-ups in the generated database.

Table 1. Different combinations of characteristic X-rays and over voltage ratios (*U*) for the synthetic $\phi(\rho r, \rho z)$ database. Entries with ✓were included.

X-ray \ U	1.2	1.5	2.0	3.0	5.0	7.0
C Kα	✓	✓	✓	✓	✓	✓
Na Kα	✓	✓	✓	✓	✓	✓
Si Kα	✓	✓	✓	✓	✓	✓
Ca Kα	✓	✓	✓	✓	✓	✓
Cr Kα	✓	✓	✓	✓	✓	✗
Cu Kα	✓	✓	✓	✓	✗	✗

Table 2. List of chosen matrix elements.

Z	Element	*Z*	Element	*Z*	Element
3	Li	15	P	25	Mn
4	Be	16	S	26	Fe
5	B	19	K	27	Co
6	C	20	Ca	28	Ni
11	Na	21	Sc	29	Cu
12	Mg	22	Ti	30	Zn
13	Al	23	V	31	Ga
14	Si	24	Cr	32	Ge

3. Results and discussions

3.1. Difficulties in modelling $\phi(\rho r, \rho z)$

Unlike $\phi(\rho z)$ distributions, $\phi(\rho r, \rho z)$ is affected by the spread of the electron beam on the sample surface. However, the distribution of electrons is hard to determine and depends on the beam energy, beam current and the type of electron gun. Hence, it is convenient to find an analytical function which works for a point beam and then convolute the function with the distribution of beam electrons as

$$\phi(\rho x, \rho y, \rho z) = \int\int \phi(\rho x - \tilde{\rho x}, \rho y - \tilde{\rho y}, \rho z) p(\tilde{\rho x}, \tilde{\rho y}) \mathrm{d}\tilde{\rho x} \mathrm{d}\tilde{\rho y} \quad (8)$$

where $p(\widetilde{\rho x}, \widetilde{\rho y})$ is the probability density function of the beam electrons.

But when using a point beam, $\phi(\rho r, \rho z)$ increases rapidly as ρr approaches 0. If we consider a small volume element near the point of beam impact, most of the ionisations will be caused by the beam electrons which have suffered minimal deviations from their original path. The number of ionisations here only depend on the size of the volume element along ρz, and consequently, the value of

$\phi(\rho r \approx 0, \rho z \approx 0)$ becomes inversely proportional to the size of the volume element along ρr. This is not favourable when computing histograms and eventually attempting to fit analytical functions onto them. Therefore, we take a cumulative version of the distribution, $\phi(\rho r, \rho z)$, accumulated along ρr which is related to $\phi(\rho r, \rho z)$ as

$$\Phi_r(\rho r, \rho z) = \int_0^{\rho r} \phi(\tilde{\rho r}, \rho z) 2\pi \tilde{\rho r} \mathrm{d}\tilde{\rho r}. \quad (9)$$

On a grid such as in Figure 1, $\Phi_r(\rho r, \rho z)$ is computed at the nodes $\widehat{\rho r}_i$ along the radial direction. $\phi(\rho r_i, \rho z_i)$ can then be calculated via numerical differentiation on the same grid as

$$\phi(\rho r_i, \rho z_j) \approx \frac{\Phi_r(\hat{\rho r}_i, \rho z_j) - \Phi_r(\hat{\rho r}_{i-1}, \rho z_j)}{2\pi \rho r_i \Delta \rho r_i}. \quad (10)$$

Finally, to avoid large variation in parameters, we use scaled coordinates ρr^* and ρz^* which are related to ρr and ρz as

$$\rho r^* = \frac{2\rho r}{D_x} \text{ and } \rho z^* = \frac{\rho z}{R_x} \quad (11)$$

where R_x is the depth of the interaction volume and D_x is the radial width (diameter) of the interaction volume. Then we define the function $\Phi_r^*(\rho r^*, \rho z^*)$ as

$$\Phi_r^*(\rho r^*, \rho z^*) = \Phi_r\left(\frac{D_x \rho r^*}{2}, R_x \rho z^*\right). \quad (12)$$

We determined R_x and D_x, individually, for all entries of the database referenced in Section 2.2, using Monte Carlo simulations, wherein, we defined R_x as the mass-depth (ρz) within which 99 % of ionisation events take place, and D_x as the mass-diameter ($2\rho r$) of an infinitely long cylinder centred at the beam, which encompasses 99 % of the ionisation events.

3.2. An analytical model for $\Phi_r^*(\rho r^*, \rho z^*)$

Examining the distributions obtained by Monte Carlo simulations, we found that the 1D distributions $\Phi_r^*(\rho r^* = const., \rho z^*)$ resemble the traditional $\phi(\rho z)$-curves, which are skewed normal-like distributions. Azzalini [14] described a general class of functions of the form

$$f(x) = 2\varphi(x)\Phi(\alpha x) \quad (13)$$

which describe such distributions, where $\varphi(x)$ is the probability density function (PDF) of some distribution which is symmetric about its mean, $\Phi(x)$ is the cumulative density function (CDF) of the same distribution and α is a tuneable parameter. We found that using the generalised normal distribution as the symmetric distribution yielded very good fits to the simulated distributions. Its PDF and CDF are given by

$$\varphi(x) = \frac{\beta}{2\sigma\Gamma(1/\beta)} \exp\left(-\left|\frac{x-\mu}{\sigma}\right|^\beta\right)$$

$$\text{and } \Phi(x) = \frac{1}{2}\left[1 + \mathrm{sign}(x-\mu)\frac{\gamma\left(\frac{1}{\beta}, \left|\frac{x-\mu}{\sigma}\right|^\beta\right)}{\Gamma(1/\beta)}\right] \quad (14)$$

respectively, where σ, μ and β are tuneable parameters, γ is the left incomplete gamma function and Γ is the gamma function.

Based on this, we define the function

$$\Phi_r^*(\rho r^*, \rho z^*) = k(\rho r^*) \exp\left(-\left|\frac{\rho z^*}{\sigma(\rho r^*)}\right|^{\beta(\rho r^*)}\right)\left[1 + \mathrm{sign}(\rho z)\frac{\gamma\left(\frac{1}{\beta(\rho r^*)}, \left|\frac{\alpha(\rho r^*)\rho z^*}{\sigma(\rho r^*)}\right|^{\beta(\rho r^*)}\right)}{\Gamma\left(\frac{1}{\beta(\rho r^*)}\right)}\right]. \quad (15)$$

This generates skewed distributions of the class described in Eq. (13) for $\Phi_r^*(\rho r^*, \rho z^*)$, where the parameters k, σ, β and α vary with ρr^*. We can now make the following observations

- By convention, $\rho z \geq 0$ since the origin of the co-ordinate system is generally placed at the point of incidence with the positive Z-axis pointing towards the material.
- $\sigma(\rho r^*)$ controls the width of the distribution. It can be restricted to be positive and greater than 0.
- $\beta(\rho r^*)$, the power inside the exponential, controls the kurtosis of the distribution. It was observed that $\beta(0) = 1$ which yields an exponential tail gives a good fit at $\rho r^* \approx 0$ and $\beta > 1$ for higher values of ρr^*.
- $\alpha(\rho z)$ controls skewness of the distribution. When α is positive the distribution skews to the positive side of the origin and vice-versa. The skewness in $\Phi_r^*(\rho r^*, \rho z^*)$ is caused by the backscattered electrons, and the distribution only skews towards the positive side of the origin. Hence, $\alpha(\rho z^*) \geq 0$.
- $k(\rho r)$ controls the scale of the distribution. Note that from Eq. (15) $\Phi_r^*(\rho r^*, \rho z^* = 0) = k(\rho r^*)$. Hence $k(\rho r^*)$ is a measure of the surface ionisation within a radius r from the electron beam, and $k(1)$ is equal to the total surface ionisation $\phi(0)$. For any ρr^*, $\Phi_r^*(\rho r^*, \rho z^* = 0) = 1 + b$, where $b \in [0,\infty)$ is a factor proportional to the total number of ionisations caused by backscattering electrons within a radius of $\rho r^* =$ const. from the origin. Hence, $k(\rho r^*) \geq 1$.

The parameters can be summarised as $\sigma(\rho r^*) \in (0,\infty)$, $\beta(\rho r^*) \in [1,\infty)$, $\alpha(\rho r^*) \in [0,\infty)$ and $k(\rho r^*) \in [1,\infty)$. When the parameters lie in the aforementioned bounds, the 'modulo' and 'sign' operators in Eq. (15) can be ignored, and the function is differentiable in the domain where $\rho r^* \geq 0$ and $\rho z^* \geq 0$.

We thus define our analytical model as

$$\Phi_r^*(\rho r^*, \rho z^*) = k(\rho r^*) \exp\left[-\left(\frac{\rho z^*}{\sigma(\rho r^*)}\right)^{\beta(\rho r^*)}\right]\left\{1 + \frac{\gamma\left[\frac{1}{\beta(\rho r^*)}, \left(\frac{\alpha(\rho r^*)\rho z^*}{\sigma(\rho r^*)}\right)^{\beta(\rho r^*)}\right]}{\Gamma\left(\frac{1}{\beta(\rho r^*)}\right)}\right\} \quad (16)$$

Further, fitting $\Phi_r^*(\rho r^* = const., \rho z^*)$ at various values of ρr^* showed that the parameter functions could be modelled with the error function (erf) as follows

$$\begin{aligned}
\sigma(\rho r^*) &= \sigma_0 + (\sigma_1 - \sigma_0)g(\rho r^*),\\
\beta(\rho r^*) &= 1 + (\beta_1 - 1)g(\rho r^*),\\
\alpha(\rho r^*) &= 1 + (\alpha_1 - 1)g(\rho r^*),\\
k(\rho r^*) &= 1 + (k_1 - 1)g(\rho r^*),\\
\text{where} \quad g(\rho r^*) &= \mathrm{erf}\left(\frac{\rho r^*}{\lambda}\right).
\end{aligned} \quad (17)$$

This results in a function with parameters λ, σ_0, σ_1, β_1, α_1 and k_1. The distribution can be represented as $\Phi_r^*(\rho r^*, \rho z^*; \lambda, \sigma_0, \sigma_1, \beta_1, \alpha_1, k_1)$. This function models the evolution of the 1D distribution $\Phi_r^*(\rho r^* = const., \rho z^*)$ from $\rho r^* = 0$ to $\rho r^* = 1$. In fact,

$$\Phi_r^*(\rho r^* = 0, \rho z^*) = \exp\left(-\frac{\rho z^*}{\sigma_0}\right)\left[1 + \frac{\gamma\left(1, \frac{\rho z^*}{\sigma_0}\right)}{\Gamma(1)}\right] \tag{18}$$

$$\Phi_r^*(\rho r^* = 1, \rho z^*) = k_1 \exp\left(-\left(\frac{\rho z^*}{\sigma_1}\right)^{\beta_1}\right)\left[1 + \frac{\gamma\left(\frac{1}{\beta_1}, \left(\frac{\alpha_1 \rho z^*}{\sigma_1}\right)^{\beta_1}\right)}{\Gamma(1/\beta_1)}\right]. \tag{19}$$

Note $\Phi_r^*(\rho r^* = 1, \rho z^*)$ is also equivalent to the traditional $\phi(\rho z)$ by definition.

3.3. Curve fitting

When given a model of the form $y = f(\mathbf{x}; \mathbf{p})$, where $\mathbf{p} = (p_1, p_2, \ldots, p_m)$ is a set of unknown m function parameters, curve fitting is the process of determining the optimum set of parameters, $\mathbf{p}^*$, which best explains a set of n observed data $(\boldsymbol{x}_1, \boldsymbol{y}_1)$, $(\boldsymbol{x}_2, \boldsymbol{y}_2)$, …, $(\boldsymbol{x}_n, \boldsymbol{y}_n)$, where y_i is the observed function value for the arguments $\boldsymbol{x}_i$ for all i. A necessary but not sufficient condition for this to be possible is that $n \geq m$. Typically, the parameters can only fall within certain ranges, which can be denoted as $\mathbf{p} \in P$. The constrained optimisation problem can then be stated in the least squares sense as follows

$$\boldsymbol{p}^* = \arg\min_{\boldsymbol{p}\in\mathcal{P}} \sum_{i=1}^{n} \epsilon\left(f(\boldsymbol{x}_i; \boldsymbol{p}) - y_i\right)^2 \tag{20}$$

where $\epsilon(r)$ is a loss function. Choosing $\epsilon(r) = r$ results in the standard least-squares problem. A good choice of the loss function can increase the robustness of the optimisation algorithm by reducing its sensitivity to data outliers and noise. We found $\epsilon(r) = 2(\sqrt{1 + |r|} - 1)$, where $|r|$ is the absolute value of r, to be a robust choice for the loss function for our problem.

Since our loss function and the model represented by Eqs. (16) and (17) are non-linear, the process of finding the optimum set of parameters requires iterative non-linear optimisation. For this, we used the Trust Region Reflective algorithm [15] as implemented in the SCIPY.OPTIMIZE.CURVE_FIT routine in the Python library, SCIPY [16]. The previously mentioned loss function can be chosen using the keyword argument LOSS='SOFT_L1'.

To make the 2D curve-fitting process even robust, we split into three distinct steps as follows:

(i) determine σ_0 by fitting $\Phi_r^* (\rho r^* = 0, \rho z^*)$,
(ii) determine σ_1, β_1, α_1 and k_1 by fitting $\Phi_r^* (\rho r^* = 1, \rho z^*)$ using the previously determined σ_0,
(iii) and finally determine λ by fitting the whole 2D function over the 2D data using all the parameters previously determined.

3.4. Results of curve-fitting

Curve fitting for the predicted function was performed for all the distributions in the generated database. From the computed root mean squared (rms) errors and visual inspection of the plots of the fitted function compared with the Monte Carlo simulated data, we can say that our model provides a good approximation for all the distributions in our database. Figure 2 shows two examples of the fits obtained. The rms errors only exceed a value of 0.1 for a very few cases. The rms errors for all the X-ray lines considered versus over-voltage ratio U and matrix atomic number Z are shown in Figure 3.

For a matrix composed of multiple elements, we recommend using an effective atomic number in place of Z. The mean atomic number $\bar{Z} = \sum_i c_i Z_i$, where c_i and Z_i are the mass fraction and atomic number of the ith element in the composition respectively, can serve as a naive estimate for the effective atomic number.

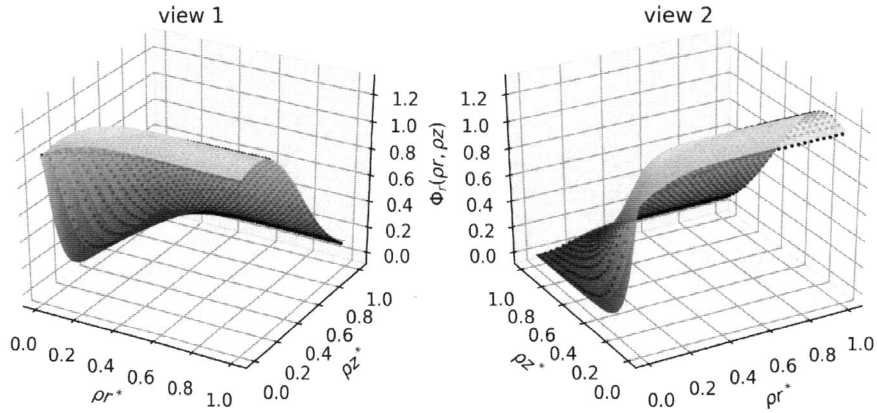

a)

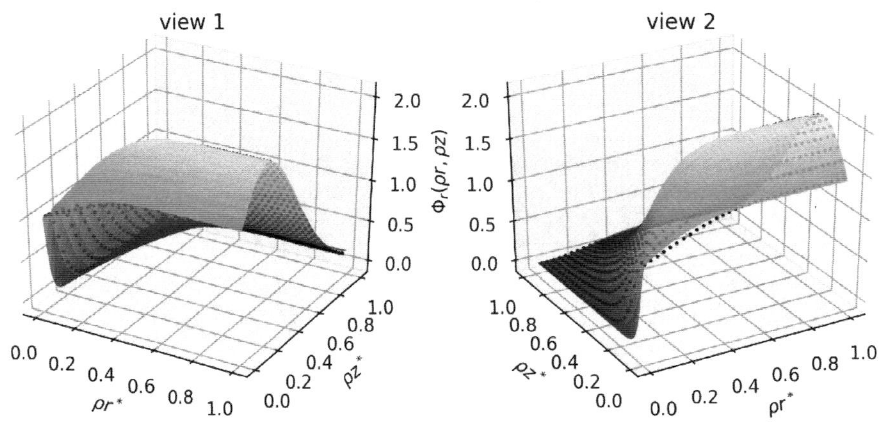

b)

Figure 2. Curve fits for Φr(ρr, ρz); points represent data generated by Monte Carlo simulations and the surface represents the fitted function. a) Si-Kα line in Ca; U = 1.5; b) Ca-Kα line in Ti; U = 3.0.

3.5. Comparison with the PAP model

Our model contains within it an approximation for $\phi(\rho z)$ as shown in Eq. (19). We compared this with the popular $\phi(\rho z)$ model, PAP [1]. The PAP model is defined based on physical parameters – the integral of the distribution F, the range of ionisation R_x, the depth of maximum ionisation R_m and the surface ionisation $\phi(0)$. Since these quantities can be calculated directly from simulated $\phi(\rho z)$ distributions, we can obtain the best fitting PAP model for any such distribution. As shown in Figure 4, our model fits better the Monte Carlo simulated data. This is because our model is more flexible and allows for different curvatures controlled by the power β inside the exponential term, whereas the PAP model is restricted to be parabolic or quadratic in nature.

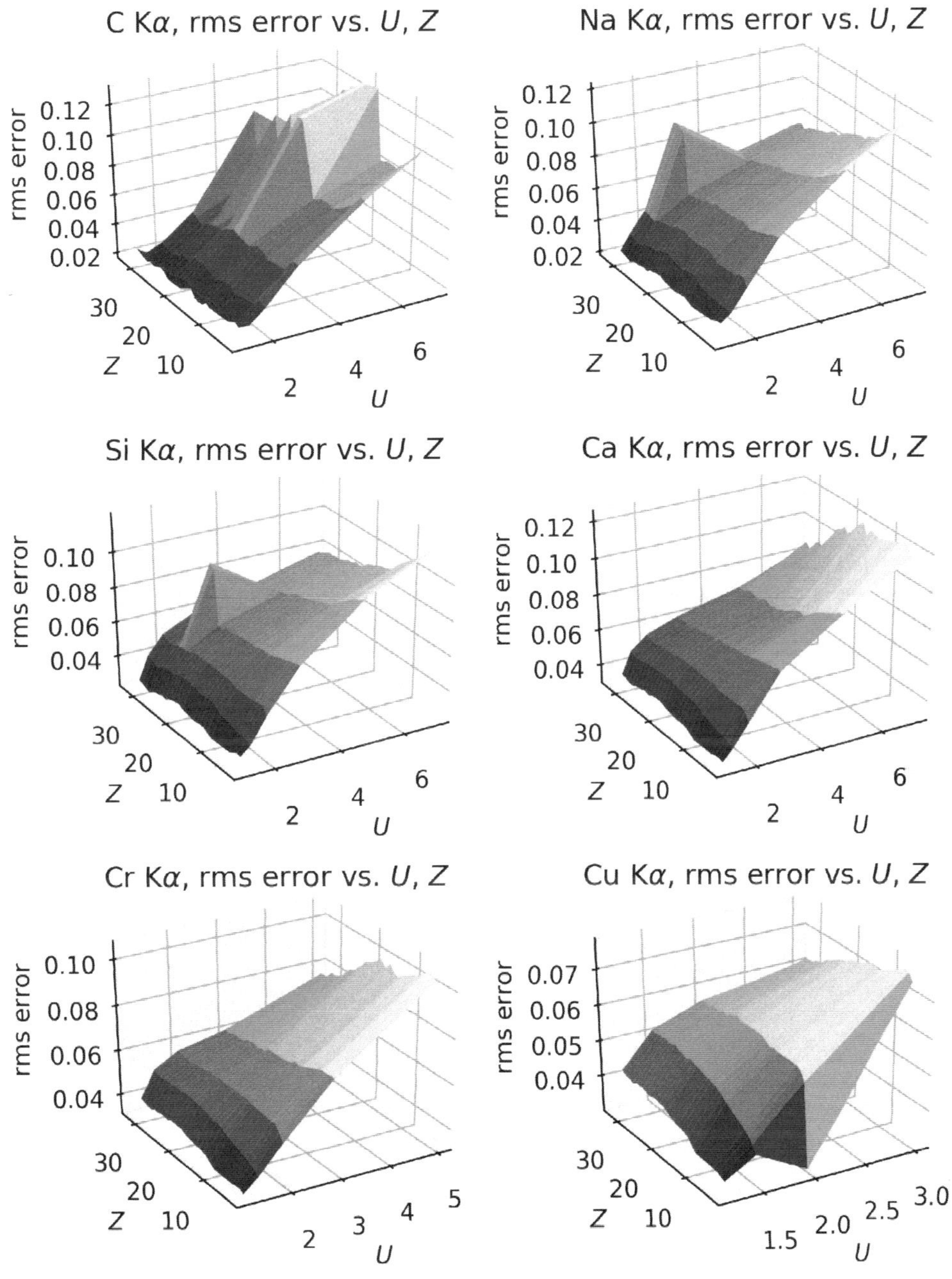

Figure 3. The root mean squared (rms) error in the obtained curve fits.

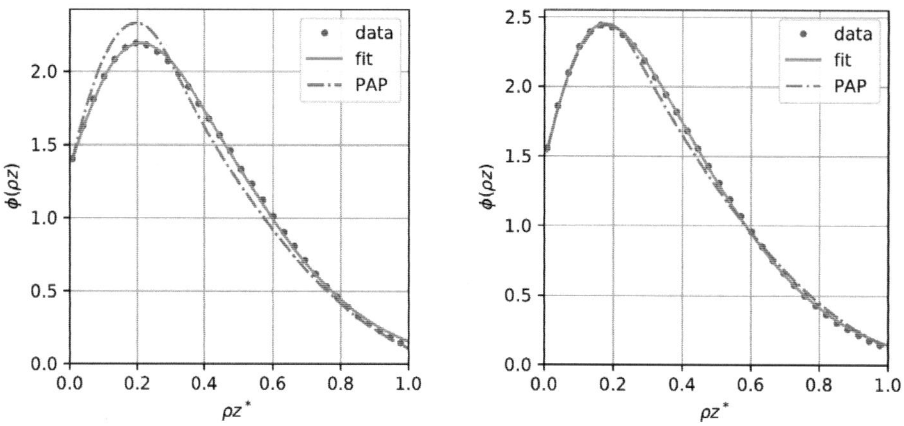

Figure 4. Comparison of $\phi(\rho z)$ obtained from our model and the PAP model with Monte Carlo simulated data.

3.6. Dependence of the parameters on over-voltage and atomic number of the matrix

We further examined the variation of the fitted function parameters obtained in Section 3.4 with respect to the simulation parameters, namely the matrix atomic number, Z, and the over-voltage ratio, U. We found that these dependencies could be well modelled with functions derived from polynomials of the form $p(x, y) = p_{00} + p_{10}x + p_{01}y + p_{20}x^2 + p_{02}y^2 + p_{11}xy$ as follows:

$$\lambda(U,Z) = p^{\lambda}{}_{00} + p^{\lambda}{}_{10}U + p^{\lambda}{}_{01}Z + p^{\lambda}{}_{20}U^2 + p^{\lambda}{}_{11}UZ \tag{21}$$

$$\frac{1}{\sigma_0(U,Z)} = p^{\sigma_0}{}_{00} + p^{\sigma_0}{}_{10}U + p^{\sigma_0}{}_{01}Z + p^{\sigma_0}{}_{20}U^2 + p^{\sigma_0}{}_{11}UZ \tag{22}$$

$$\frac{1}{\sigma_1(U,Z)} = \beta_1(U,Z)\,(p^{\sigma_1}{}_{00} + p^{\sigma_1}{}_{10}U + p^{\sigma_1}{}_{01}Z) \tag{23}$$

$$\frac{1}{\beta_1(U,Z)} = p^{\beta_1}{}_{00} + p^{\beta_1}{}_{10}U + p^{\beta_1}{}_{01}Z + p^{\beta_1}{}_{20}U^2 + p^{\beta_1}{}_{11}UZ \tag{24}$$

$$\alpha_1(U,Z) = p^{\alpha_1}{}_{00} + p^{\alpha_1}{}_{10}U + p^{\alpha_1}{}_{01}Z + p^{\alpha_1}{}_{20}U^2 + p^{\alpha_1}{}_{11}UZ \tag{25}$$

$$k_1(U,Z) = p^{k_1}{}_{00} + p^{k_1}{}_{10}U + p^{k_1}{}_{01}Z + p^{k_1}{}_{20}U^2 + p^{k_1}{}_{11}UZ \tag{26}$$

For the parameters obtained in section 3.4, we fitted equations (21) to (26) and the results are tabulated in Table 3. Further, as an example, Figure 5 shows the obtained fits for the parameters of Cr-Kα line. Since $\phi(\rho r, \rho z)$ is a distribution of ionisation probability, it is influenced by the X-ray line being measured, in addition to the electron beam energy and the matrix or sample composition. This necessitates the use of individual sets of parameters for each X-ray line, as seen in Table 3.

Table 3. Results of curve fitting for the parameters obtained in section 3.4 as functions of the over-voltage ratio U and atomic number Z of the matrix element.

X-ray / param.	C K-α	Na K-α	Si K-α	Ca K-α	Cr K-α	Cu K-α
$p^{\lambda}{}_{00}$	0.2152	0.2689	0.2793	0.2871	0.2704	0.2502
$p^{\lambda}{}_{10}$	0.0628	0.06632	0.06185	0.05712	0.05084	0.03492
$p^{\lambda}{}_{01}$	-0.01137	-0.004052	-0.0017	0.0005057	0.002787	0.005897
$p^{\lambda}{}_{20}$	-0.0119	-0.02148	-0.01971	-0.01508	-0.009105	-0.002875
$p^{\lambda}{}_{11}$	-0.0005675	0.0001905	-0.0007542	-0.004203	-0.003621	-0.002234
$p^{\sigma_0}{}_{00}$	10.88	29.09	43.24	70.04	77.19	76.02
$p^{\sigma_0}{}_{10}$	5.545	15.93	22.91	31.06	30.38	19.09
$p^{\sigma_0}{}_{01}$	-0.8858	-1.739	-1.733	2.007	0.966	1.432
$p^{\sigma_0}{}_{20}$	-0.5978	-1.94	-3.345	-7.382	-9.066	-5.986
$p^{\sigma_0}{}_{11}$	-0.6198	-1.022	-0.8613	2.968	2.984	3.025
$p^{\sigma_1}{}_{00}$	3.152	3.297	3.31	3.402	3.431	3.452
$p^{\sigma_1}{}_{10}$	0.03859	0.04248	0.005444	0.04325	0.000894	0
$p^{\sigma_1}{}_{01}$	-0.04529	-0.0743	-0.08516	-0.1023	-0.1082	-0.1218
$p^{\beta_1}{}_{00}$	0.5847	0.5313	0.5192	0.5289	0.5324	0.5362
$p^{\beta_1}{}_{10}$	-0.02038	-0.04227	-0.02896	-0.01926	-0.00466	0.007629
$p^{\beta_1}{}_{01}$	0.04023	0.05032	0.04999	0.04966	0.05157	0.05639
$p^{\beta_1}{}_{20}$	-0.02318	0.005762	0.003811	-0.02527	-0.02491	-0.02426
$p^{\beta_1}{}_{11}$	-0.0002961	-0.007288	-0.007887	-0.006772	-0.008086	-0.008156
$p^{\alpha_1}{}_{00}$	1.665	1.696	1.822	1.839	1.782	1.559
$p^{\alpha_1}{}_{10}$	0.7576	0.6344	0.6709	0.8395	0.8429	0.6898
$p^{\alpha_1}{}_{01}$	0.2317	0.2349	0.2739	0.3857	0.4103	0.4269
$p^{\alpha_1}{}_{20}$	-0.3451	-0.2102	-0.2485	0.05064	-0.0237	0.01342
$p^{\alpha_1}{}_{11}$	0.09934	0.02213	0.0544	0.1314	0.1157	0.08212
$p^{k_1}{}_{00}$	1.285	1.417	1.438	1.498	1.424	1.335
$p^{k_1}{}_{10}$	0.1874	0.2001	0.2	0.2	0.1707	0.09984
$p^{k_1}{}_{01}$	0.009355	0.06924	0.08309	0.09207	0.08387	0.07635
$p^{k_1}{}_{20}$	-0.03206	-0.06352	-0.0593	-0.1062	-0.08182	-0.04631
$p^{k_1}{}_{11}$	0.01749	0.03971	0.03238	0.02966	0.02984	0.03078

4. Summary and outlook

We showed that the three-dimensional ionisation probability distribution can effectively be modelled with a two-dimensional model for $\phi(\rho r, \rho z)$. We then derived an analytical model for $\Phi_r^*(\rho r^*, \rho z^*)$, which is $\phi(\rho r, \rho z)$ accumulated along ρr and expressed in terms of normalised coordinates as explained in sections 3.1 and 3.2. We fitted the model to a comprehensive database of distributions generated by Monte Carlo simulations. The fits reveal the versatility of our model in describing the various forms of ionisation distributions that can result from different experimental conditions. The parameters obtained from the fits were further modelled as functions of the characteristic X-ray line, over-voltage ratio, U, and the atomic number of the matrix, Z.

When used as it is, the model can be used for samples that are homogeneous or where the heterogeneities have negligible influence on the ionisation distribution. But in order to extend to more general heterogeneous samples, an appropriate weighting procedure has to be developed. This will be a subject of future work. Further, we envision to optimise the number of parameters and the means of calculating them.

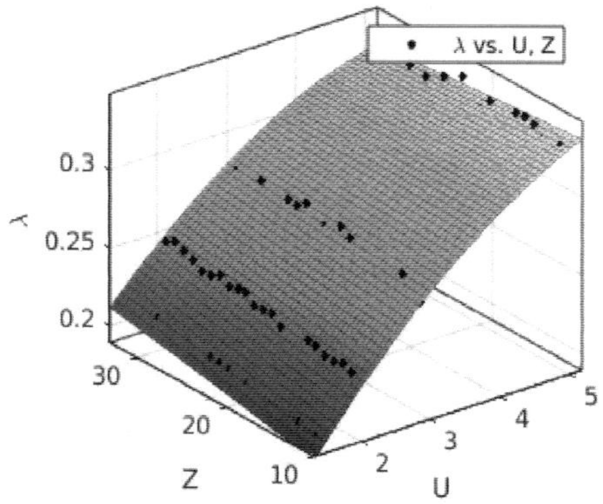

a) Parameter λ.

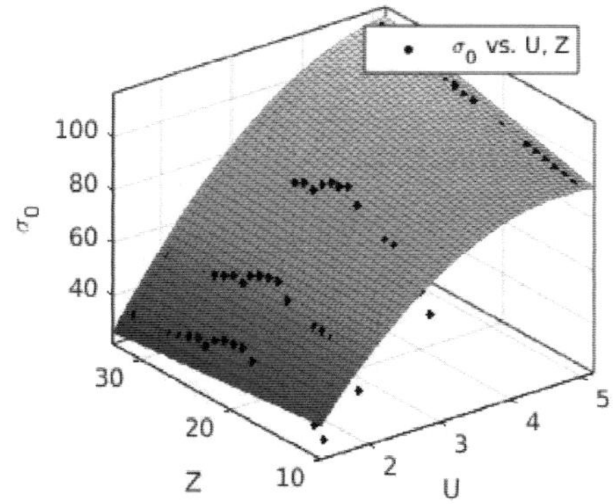

b) Parameter σ_0.

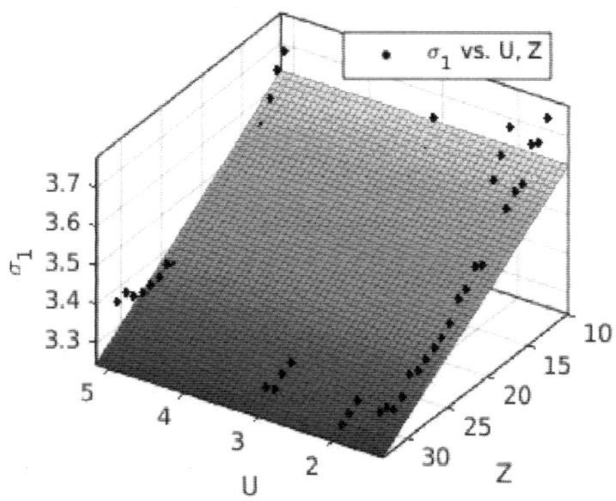

c) Parameter σ_1.

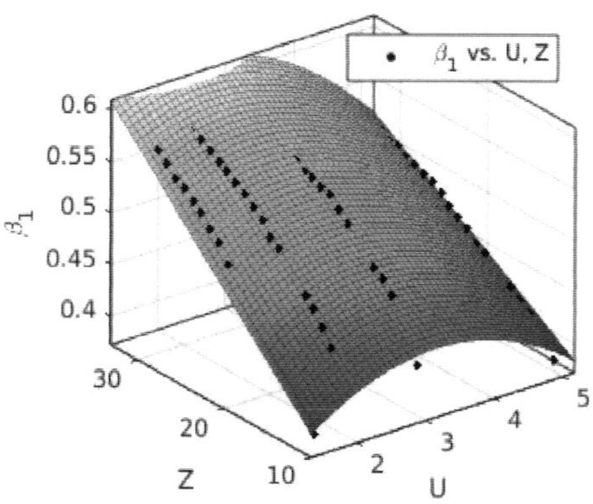

d) Parameter β_1.

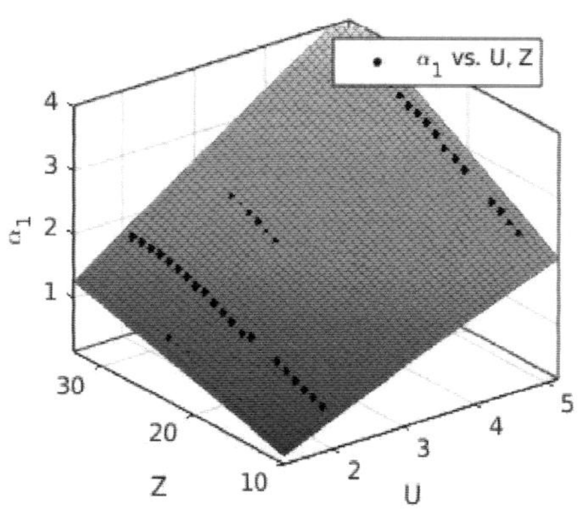

e) Parameter α_1.

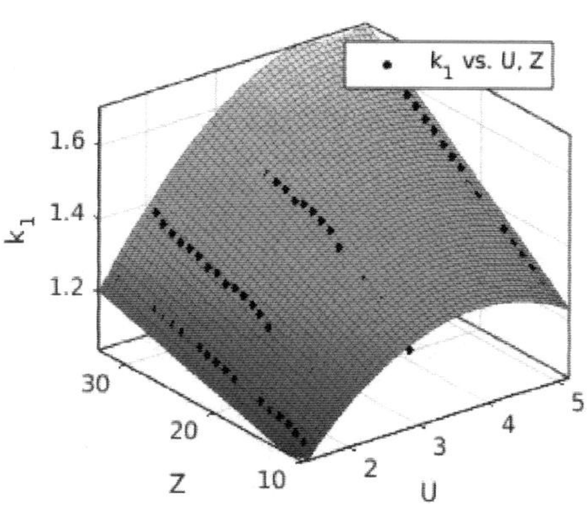

f) Parameter k_1.

Figure 5. Curve fitting results for the model parameters for Cr Kα X-ray line as functions of over-voltage ratio U and atomic number of the matrix element Z.

5. References

[1] Pouchou J L and Pichoir F 1991 in: *Electron probe quantitation.* (Heinrich K F J and Newbury D E; Eds.) (Boston, MA: Springer US) 31-75
[2] Waldo R A 1988 in: *Microbeam analysis.* (Newbury D E; Ed.) (San Francisco: San Francisco Press, Inc) 310-314
[3] Llovet X and Merlet C 2009 *Microsc. Microanal.* **16** 21-32
[4] Wagner H W, Werner W S M, Störi H and Richardson L M 2001 *Nucl. Instrum. Methods Phys. Res. Sec. B: Beam Interact. Mater. Atoms* **184** 450-457
[5] Richter S, Pinard P T, Mevemkamp N and Torrilhon M 2013 *Microsc. Microanal.* **19** 1296-1297
[6] Yuan Y, Demers H, Brodusch N, Wang X and Gauvin R 2020 *Ultramicroscopy* **219** 113117
[7] Bunger J 2021 *Three-dimensional modelling of X-ray emission in electron probe microanalysis based on deterministic transport equations.* Ph.D. thesis. (Aachen, Germany: RWTH Aachen University)
[8] Bünger J, Richter S and Torrilhon M 2022 *Microsc. Microanal.* **28** 454-468
[9] Claus T, Bünger J and Torrilhon M 2021 *Math. Computational Appl.* **26** 51
[10] Castaing R 1951 *Application of electron probes to metallographic analysis.* Ph.D. thesis. (Paris, France: University of Paris)
[11] Castaing R and Descamps J 1955 *J. Physique Radium* **16** 304-317
[12] Karduck P and Rehbach W 1991 in: *Electron probe quantitation.* (Heinrich K F J and Newbury D E; Eds.) (Boston, MA: Springer US) 191-217
[13] Ritchie N W M 2005 *Surface Interface Anal.* **37** 1006-1011
[14] Azzalini A 1985 *Scand. J. Statistics* **12** 171-178
[15] Branch M A, Coleman T F and Li Y 1999 *SIAM J. Sci. Computing* **21** 1-23
[16] Virtanen P, *et al.* 2020 *Nature Methods* **17** 261-272

Measurements of the quantitative analytical depth resolution at evaporated gold layers with the FEG-EPMA JEOL JXA-8530F

D Berger and J Nissen

Technische Universität Berlin, Zentraleinrichtung Elektronenmikroskopie ZELMI, Straße des 17. Juni 135, 10623 Berlin, Germany

E-mail: dirk.berger@tu-berlin.de, joerg.nissen@tu-berlin.de

Abstract. This work concludes our systematic investigations of the spatial analytical resolution at different primary electron energies and specimens with different atomic numbers [1-4]. The aim of the entire study was to determine the qualitative and especially the quantitative analytical spatial resolutions that can actually be achieved in practise. For this purpose, we prepared test specimen by evaporating thin metallic layers (Al, Ag and Au) with various thicknesses on top of non-fluorescent substrates. To determine the depth resolution, the layers were analysed from the top. The measurements were carried out using a field emission electron microprobe, but the results are also applicable to SEM-EDS and SEM-WDS.
This work now evaluates the quantitative analytical depth resolution measured with electron energies from 4 to 15 keV on eight gold layer specimens with different layer thicknesses from 30 to 1,030 nm. By varying the primary electron energy E_0, we determined the minimum value of E_0 for which the depth of the signal source volume is identical to the layer thickness. This value is defined as the depth resolution, which is identical to the ionisation range of the atomic shell under consideration. With respect to the excitation energy of the characteristic X-rays, the depth resolution for the quantitative element analysis for Au (Mα at 4 kV) is found to be below 50 nm. It should be noted that the applied electron energies fulfil the 2-times overvoltage criteria resulting in reasonable X-ray signals with good signal-to-noise ratio.
The experimental results were cross-checked with simulations and calculations. While we found a good agreement with Monte Carlo simulations, the well-known Castaing-formula predicts smaller values for all primary electron energies. The deviation increases for decreasing energies. Since we regard test specimens without any disturbing secondary fluorescence, the measurements presented in this paper are very helpful for further improvements of calculations, MC simulations and other approximations.

1. Introduction
The element analysis of thin surface layers is of great interest for scientific research and technological development across various fields like metallic coatings on metals and glasses, hard coatings on tools, surface reactions and separation in geology and soil science, electroplating, catalysts and fuel cells. Electron probe techniques such as scanning electron microscopy energy-dispersive X-ray spectrometry (SEM-EDS), scanning electron microscopy wavelength-dispersive X-ray spectrometry (SEM-WDS), or electron probe microanalysis (EPMA) with low-kV electron optics are commonly employed for these measurements due to their ability to provide high spatial resolution.

The quantitative element composition analysis of thin surface layers requires special considerations to ensure accurate results. Either the measurement parameters (E_0 and X-ray line) should be set in a way that the ionisation range (over which X-rays are generated) is smaller than the layer thickness, or sophisticated thin film software packages (e.g., STRATAGEM [5], …), calculations (e.g., Llovet [6]) or simulations (e.g., CASINO [7], PENEPMA [8], BADGERFILM [9]) are utilised to analyse thin surface layers. In all cases, a detailed understanding of the electron beam-specimen interaction with respect to the depth resolution is crucial. This understanding helps both in selecting appropriate measurement parameters and in the validation of the results obtained by thin film software, calculations and simulations.

Therefore, we have started a systematic evaluation of the spatial resolution in WDS element analysis for specimens with varying atomic numbers. First of all, we differentiate between quantitative and qualitative resolution, the latter only confirming the presence or absence of specific elements while the first determines the percentage of elements within the specimen. Secondly, the spatial resolution evaluation is divided into lateral (typically the x-/y-direction) [2, 3] and depth (z-direction) resolutions [4]. The measurements of the quantitative depth resolution are extended to pure gold layers in the present paper to complete the systematic evaluation.

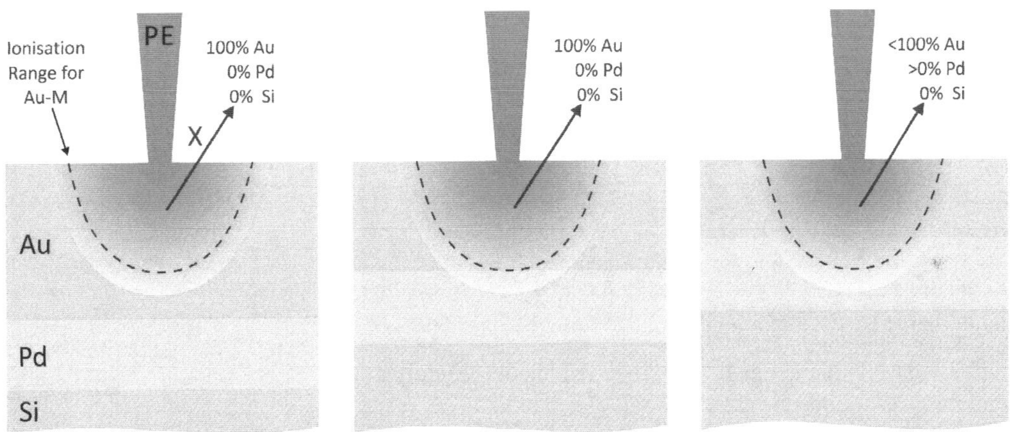

Figure 1. Measurement procedure to determine the depth resolutions at given beam conditions (E_0, I_0) from specimens with different gold layer thicknesses for each of which a quantitative WDS analysis of Au, Pd and Si is acquired; the Pd intermediate layer reduces secondary fluorescence effects from Si.

2. Experimental and results

2.1. Measurement scheme and test specimens

The quantitative depth resolution for a given primary electron energy is defined as the thinnest surface layer thickness for which the quantitative element analysis using WDS detects 100 % of the surface layer material, which in our case is pure gold, see Figure 1. Obviously, the thus defined quantitative depth resolution is closely related to the ionisation range. If the layer thickness is larger compared to the ionisation range (left part in Figure 1), the quantification result gives 100 % Au as well. If the layer thickness is less than the ionisation range (right part of Figure 1), we obtain less than 100 % Au but some percentage of the underlying material, which in our case is palladium.

To evaluate the depth resolution in dependence on the energy of the primary electrons, we prepared test specimens with gold layer thicknesses of 1,030, 250, 170, 120, 90, 75, 50 and 30 nm. As microscopists prefer to have integer numbers of the electron energies (4, 5, 6, 7, 8 keV), we cannot measure the exact numerical value of the depth resolution. Therefore, our experimental results are given as number ranges between the thinnest layer with 100 % Au and the next thinner layer with a little bit less than 100 %, see Table 1.

Table 1. Results of quantitative gold layer measurements in wt% Au.

Au layer thickness [nm]	15 keV	8 keV	7 keV	6 keV	5 keV	4 keV
1030 (Reference)	100,00	100,00	100,00	100,00	100,00	100,00
250	95,92	99,93	99,95	100,10	100,02	99,91
170	81,37	99,87	99,91	100,08	100,02	99,99
120	/	97,39	99,53	99,92	99,96	100,01
90	/	/	/	99,24	99,88	100,05
75	/	/	/	/	98,82	99,99
50	/	/	/	/	93,38	99,59
30	/	/	/	/	/	88,86

It should be noted that the "100 %" is only a figure of merit. In electron microprobe analysis, a quantification result in the range of 99.5 to 100.5 % is typically well satisfying.

In addition, the depth resolution depends strongly on the detected X-ray line. Our systematic evaluation is limited to the low-kV analysis and excludes the low overvoltage analysis. Therefore, we only consider the Au-Mα line. The minimum primary electron energy is 4 keV, which is twice the Au-M critical ionisation energy of 2.307 keV [10].

The Pd interlayer (Figure 2) offers the possibility to double check the findings: The decrease in Au concentration coincides with the increase of the Pd concentration. Pd X-rays are only excited by electrons and not by Au-fluorescence, i.e., the Pd signal is a measure of the layer thickness as well. However, Bremsstrahlung X-rays could excite Pd-L lines, which could then activate Au-M fluorescence and distort the results. We expect these continuum fluorescence effects to be negligible within the measurement error of our quantized (in primary electron energy and film thicknesses) approach. If there were no Pd interlayer, we would always observe an Au – Si fluorescence, i.e., the Si signal would not be a direct measure of the depth resolution.

The test specimens were prepared by vapour deposition on silicon wafer substrates. The layer thickness is determined from calibrated SEM images of cross sections prepared by broad Ar-ion beam milling (JEOL Cross Section Polisher IB-19520CCP), see Figure 2.

The density of the Au-layers was determined by the weighting method [11 to be (19.29 +/- 0.01) g/cm^3. This density value is close to the bulk material density indicating that the evaporated layers are not porous. Previously we have found that sputtered layers may be porous or have a different crystal structure and thus have a different density compared to bulk samples.

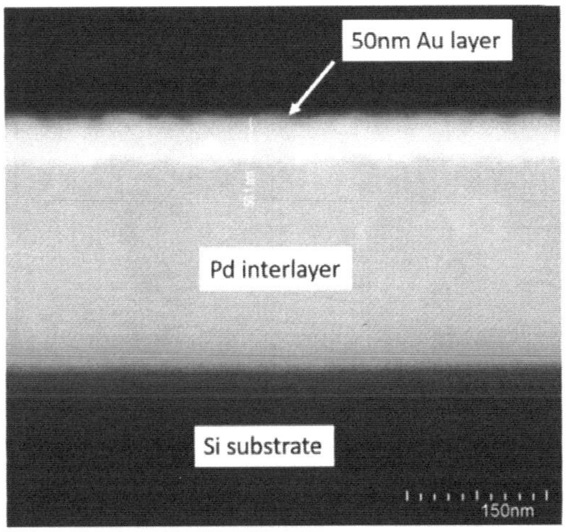

Figure 2. SEM image of a cross-section of a 50 nm gold layer with a 250 nm palladium interlayer; Cross-section prepared by broad ion beam milling.

2.2. Measured quantitative depth resolution

The measurements are performed with a JEOL JXA-8530F field emission electron microprobe. The Au-Mα, the Pd-Lα and the Si-Kα signals are recorded for different primary electron energies (4 ... 15 kV) and for all eight samples (thicknesses 30 ... 1,030 nm) with the PET-H (Au-Mα), PET-H (Pd-Lα) and TAP (Si-Kα) crystals respectively. Table 1 summarises the results of the element quantification (including ZAF correction) of gold.

The 1,030 nm Au layer serves as reference standard for the element quantification. Its thickness significantly exceeds the estimated depth of the electron interaction volume in Au of about 115 nm at 7 keV and 45 nm at 4 kV (see Fig. 3). The beam current was 10 nA and the acquisition time was 20 s for the peak and 10 s for the background.

The values in the table are the averages of 10 measurements in weight percent of gold (wt% Au). A quantification result of (100.0 +/- 0.5) % indicates that the source volume of the emitted X-rays is completely inside the Au layer. The 50 nm layer is the thinnest that can be resolved (99.59 wt% Au). Thus, the best quantitative depth resolution in the range of 30 to 50 nm is achieved for gold at 4 keV electron energy. From our measurements it might be concluded only that the depth resolution is below 50 nm and above 30 nm. A rough interpolation indicates a value of about 45 nm. Additional test specimens would be required to determine the depth resolution more accurately. So far, we specify the measured depth resolution to be at least 50 nm for Au-Mα at 4 kV.

2.3. Monte-Carlo simulation of the quantitative depth resolution

For the reason of comparison Monte Carlo simulations are performed using the CASINO software (version 2.48 [7]) for the same primary electron energies. Each of the 10,000 electron paths is stopped at the critical ionisation energy of 2.307 keV for Au M. Figure 3 shows the resulting plots of the electron paths of primary electrons (blue) and backscattered electrons (red). Our goal to analyse the quantitative depth resolution within a confidence interval of 99.5 to 100.5 % is taken into account by neglecting the deepest 50 paths. The depth of the next deepest path is taken as the depth resolution, as indicated in Figure 3.

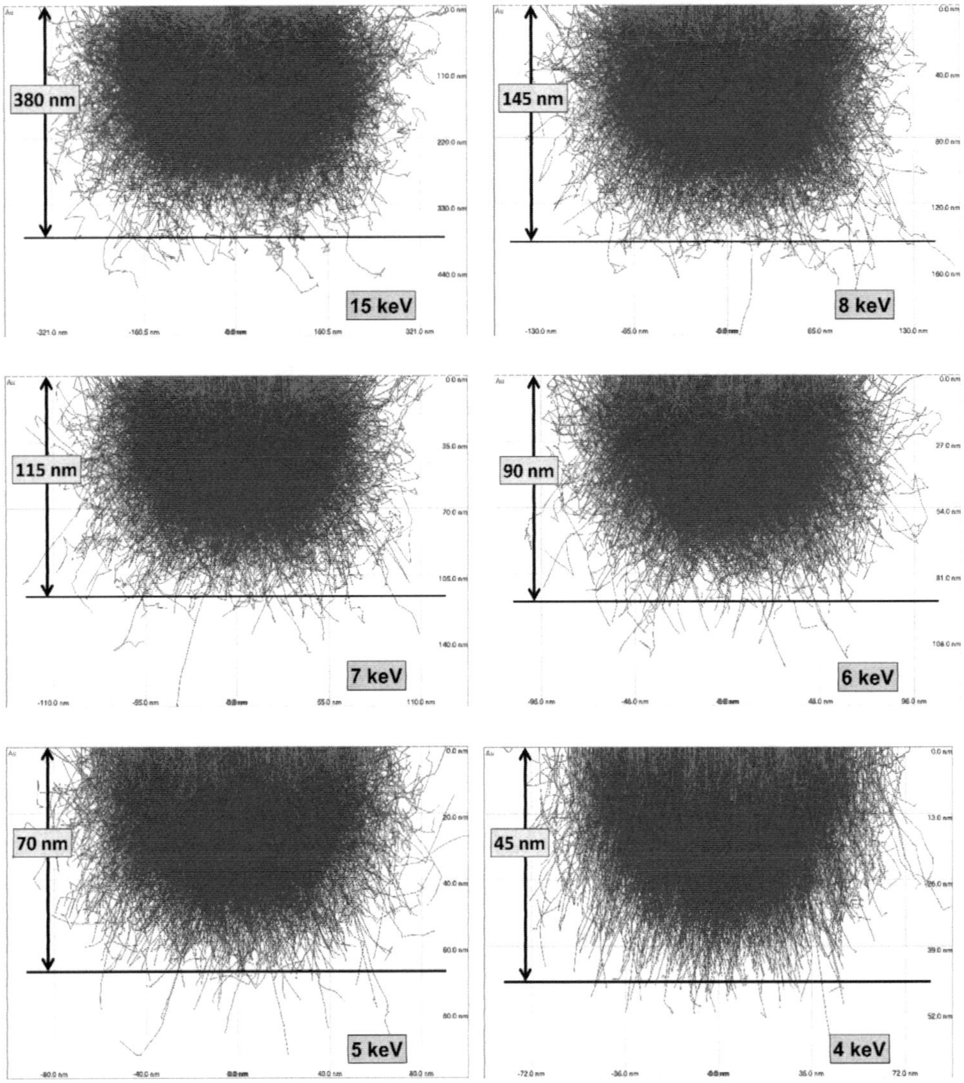

Figure 3. Quantitative depth resolution derived from Monte Carlo simulations (CASINO 2.48) of electron paths in Au for different primary electron energies.

The simulated $\phi(\rho z)$-curves in Figure 4 show that the self-absorption of Au-Mα in Au is insignificant for the low-kV excitation with primary electron energies up to 8 keV. Therefore, the range of the electron paths with an energy above the critical ionisation energy is a good measure for the ionisation range of emitted X-rays. The negligibility of self- and secondary fluorescence of our test specimens is considered to be a great advantage for further validation of formulas for calculating the ionisation range of electrons or of Monte Carlo simulations at low primary electron energies.

IOP Conf. Series: Materials Science and Engineering **1324** (2025) 012002 doi:10.1088/1757-899X/1324/1/012002

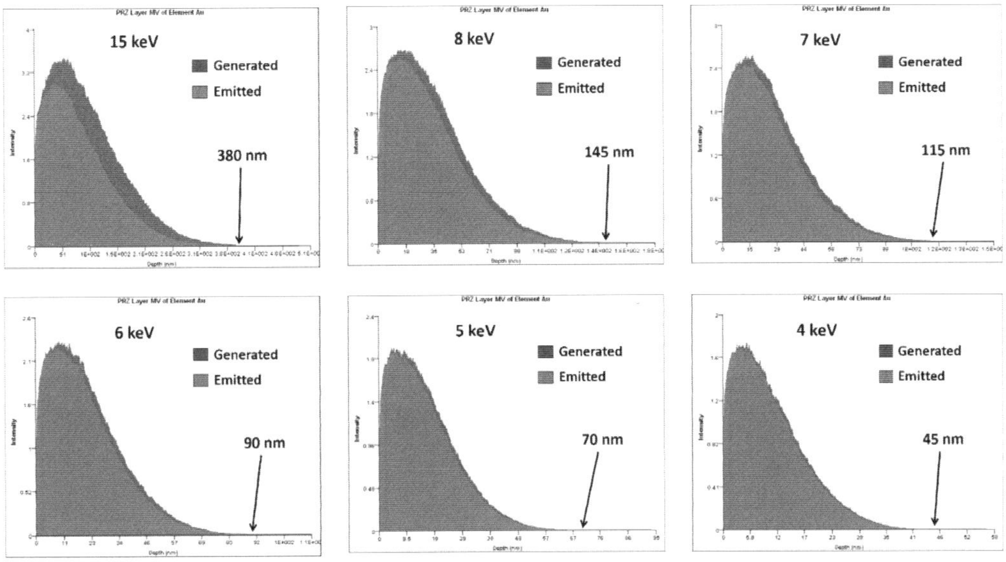

Figure 4. Estimation of self-fluorescence effects of Au-Mα X-rays generated in Au for different primary electron energies from simulated $\phi(\rho z)$-curves.

2.4. Calculation of the quantitative depth resolution

From Figure 1 we have deduced that the quantitative depth resolution is a good measure of the ionisation range and vice versa. The latter can be calculated for comparison from Castaing's famous equation [12]:

$$R_x = \frac{33A}{\rho \cdot Z}\left(E_0^{1.7} - E_i^{1.7}\right), \quad (1)$$

where R_x is the range of the electrons before slowing down to an energy equal to the critical ionisation energy of the considered electron shell [6], A is the atomic mass, ρ is the measured (see Section 2.1) density, Z is the atomic number, E_0 is the energy of primary electrons and E_i is the ionisation energy of the Au-M shell.

Since the self-absorption of Au-Mα in Au is proven to be very small (see $\phi(\rho z)$-curves), absorption effects do not need to be taken into account. The calculated values are given in the last column of Table 2.

2.5. Comparison of measured, simulated and calculated values of the quantitative depth resolution

Finally the measured values (Table 1), simulated values (Figure 3) and calculated values (Eq. 1) are summarised in Table 2 and plotted in Figure 5. Obviously, the depth resolution decreases significantly with decreasing primary electron energy due to the reduction of the range of the scattered electrons. Measured and simulated values coincide quite well over the whole range of primary electron energies including the indicated saddle point between 6 and 7 keV and the drop below 5 keV. However, the reasons for both effects are not clear and, to our knowledge, are not described in the literature. In addition we have not previously found a similar indication of a saddle point for aluminium and silver specimens [4].

Table 2. Results of measured, simulated and calculated analytical depth resolutions.

Electron energy	Measured depth resolution range [nm]	MC simulation depth resolution [nm]	Calculated depth resolution [nm]
15 keV	250-1030	380	408
8 keV	120-170	145	129
7 keV	90-120	115	99
6 keV	90-120	90	72
5 keV	75-90	70	48
4 keV	30-50	45	27

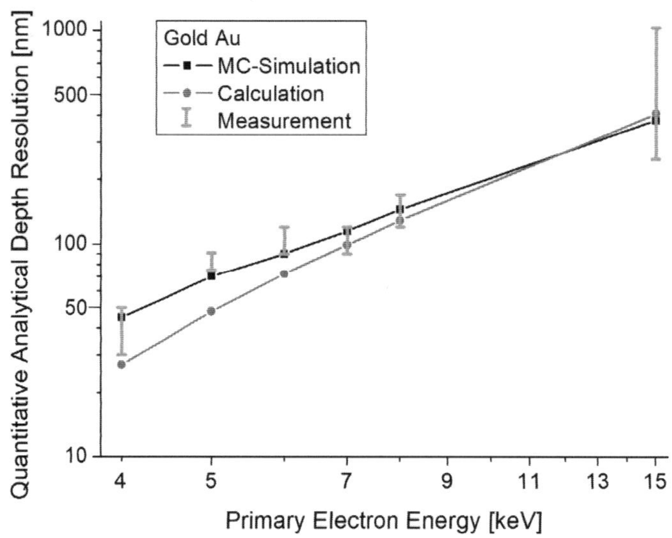

Figure 5. Plotted results from Table 2.

There are larger discrepancies between measured and calculated values for primary electron energies below 7 keV, similar to the previous findings for silver specimens in both lateral and depth resolutions [3-4]. Unfortunately, the reasons for this are still unclear. The hypotheses put forward in the earlier publication are still neither confirmed nor refuted. We hope that the measurements of our present investigation will help to resolve these contradictions in the future.

3. Conclusion

In this paper, we successfully confirm that a previously developed experimental method for determining the quantitative analytical depth resolution [4] delivers reasonable results even at specimens with a high (mean average) atomic number. The presented proof is performed with different primary electron energies on thin gold films of different thicknesses. Hereby, the depth resolution is defined as the ionisation range of the Au-M shell, which is commonly used for element analysis with high spatial resolution and high overvoltage ratio, i.e., high signal-to-noise ratio of the analysed X-ray line.

IOP Conf. Series: Materials Science and Engineering **1324** (2025) 012002 doi:10.1088/1757-899X/1324/1/012002

The applied experimental method with discrete electron energies and layer thicknesses implies that the depth resolution is obtained as a range. Here, we found that the minimum depth resolution is in the range of 30 to 50 nm at 4 keV. This number is in good agreement with previously obtained values for different elements [4] and the lateral analytical resolution [2, 3].

The experimental results are in agreement with Monte-Carlo simulations for incident electron energies in the range from 4 to 15 keV, but clearly differ from the prediction of the well-known Castaing equation (Eq. 1) which predicts lower values especially at low electron energies. The depth resolution measurements presented here are valuable for further improvements of both simulations and model-based approximations, as well as of improved calculation equations. Our specimens provide a simple test case where secondary fluorescence is negligible.

Acknowledgement

The financial support by the Europäische Fonds für Regionale Entwicklung (EFRE) within the projects "Nano-Analytiklabor" and “Nano-Werkbank” is gratefully acknowledged.

References

[1] Berger D and Nissen J 2014 *IOP Conf. Ser.: Mater. Sci. Eng.* **55** 012002
[2] Berger D and Nissen J 2016 *IOP Conf. Ser.: Mater. Sci. Eng.* **109** 012001
[3] Berger D and Nissen J 2018 *IOP Conf. Ser.: Mater. Sci. Eng.* **304** 012002
[4] Berger D and Nissen J 2020 *IOP Conf. Ser.: Mater. Sci. Eng.* **891** 012004
[5] *Stratagem Version 6.7*, SAMx, 1554 route de la Roquette, 06670 Levens, France
[6] Merlet C and Llovet X 2012 *IOP Conf. Ser.: Mater. Sci. Eng.* **32** 012016
[7] Drouin D, *et al.* 2007 *Scanning* **29**(3) 92-101
[8] Salvat F, Llovet X, Fernández-Varea J M and Sempau J 2006 *Microchim. Acta* **155** 67
[9] Moy A and Fournelle J H 2021 *Micros. Microanal.* **27** 266
[10] Carpenter P K and Jolliff B L 2015 *Microsc. Microanal.* **21** (Suppl. 3) 1434-1444
[11] Berger D and Galbert F 2011 in: *Conf. Proceedings MC2011.* (Kiel, Germany) ISBN 978-3-00-033910-3 IM7.P208
[12] Castaing R 1960 *Adv. Electronics Electron Phys.* **13** 317

STEM-CL and related spectroscopies using a high numerical aperture mirror

M Bézard[1], J Baaboura[1], Y Auad[1], N Varkentina[1], N Bonnet[1], S Y Woo[1], M Tencé[1], J-D Blazit[1], X Li[1], S Meuret[2], F Houdellier[2], H Lourenço-Martins[2], A Arbouet[2], O Stéphan[1], L H G Tizei[1] and M Kociak[1]

[1] Université Paris-Saclay, CNRS, Laboratoire de Physique des Solides, 91405, Orsay, France
[2] CEMES, UPR 8011, CNRS-Université de Toulouse, 29 Rue Jeanne Marvig, P.O. Box 94347, 31055 Toulouse, France

E-mail: mathieu.kociak@universite-paris-saclay.fr

Abstract. This conference proceeding reports on the usage and applications of a high numerical aperture mirror for cathodoluminescence and other photon-based spectroscopies in a scanning transmission electron microscope.

1. Introduction

In 2015, one of us (MK) was invited to give a lecture on STEM-CL at the 2015 EMAS workshop. Indeed, the technique was then undergoing a revival, mainly driven by the pioneer group in the field (Prof. Yamamoto's team, now Sannomiya's) and the one at LPS in Orsay. As we noted at the time, this revival also accompanied that of cathodoluminescence in the SEM. In this lecture, we presented the salient results of a nascent field, and from this course emerged a few years later a journal article on STEM-CL [1].

The advances obtained in the Orsay STEM group were essentially linked to the possibility of driving, in the reduced and constrained space of the TEM objective lens gap, a high numerical aperture mirror with great precision and reproducibility in the positioning. The common point of the experiments described in this paper and obtained in the teams of the LPS at Orsay and the CEMES in Toulouse is also this mirror, to which innovative technologies are attached. It is worth noting that several authors have financial interests with the Attolight company, which sells now the mirror as a commercial product.

Figure 1 shows schematically a STEM with a gun (which may or may not be pulsed), possibly a monochromator, a mirror to detect or inject light, and an EELS spectrometer. Depending on its mode of operation, several signals can be detected or injected, with or without time-resolution and with or without synchronisation. The main part of this proceeding will concern such correlative, coincident, or synchronised experiments, which is emphasised in Figures 1 and 2.

7 years back ago, the results concerning the Orsay group presented at EMAS in 2015 were those shown in Figure 1 and obtained on a VG HB501 microscope, which is now more than 40 years old. Despite its advanced age, results are still being obtained on the VG. But we will mainly present results obtained on the FEMTOTEM microscope from Toulouse (Hitachi microscope on which is mounted a unique pulsed field effect gun) and the CHROMATEM monochromatic microscope (customised version of the NION HERMES200) and summarised on Figure 2.

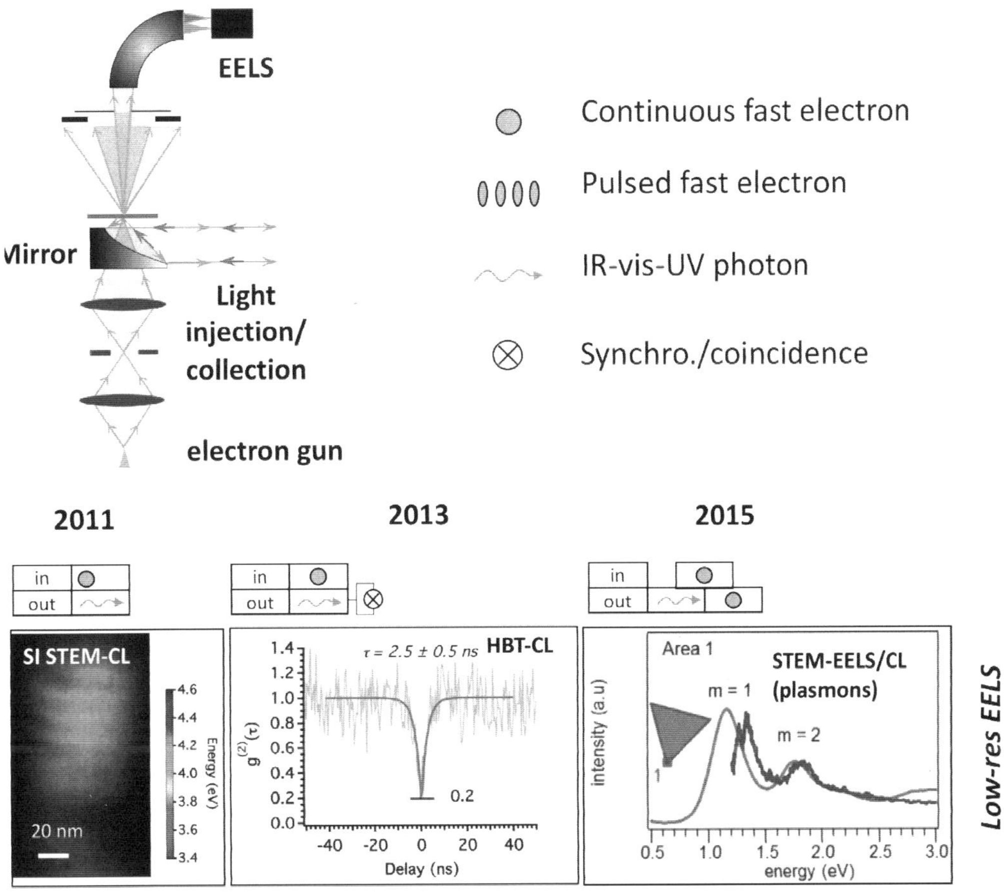

Figure 1. EELS and CL based spectroscopies in (S)TEM, and how they can be combined together. Top, left: Schematics of a (S)TEM fitted with a gun (optionally pulsed), a monochromator (optional), a high numerical aperture mirror for light injection and detection, and an EELS spectrometer. Top, right: pictogram convention for Figure 1 (bottom) and Figure 2. Bottom: some milestones experiments (from left to right [2, 3, 4]), at the time of the lecture on STEM-CL at the 2015 edition of EMAS, obtained on a continuous and non-monochromated VG microscope. SI: spectral-imaging. HBT: Hanbury-Brown and Twiss (intensity) interferometer. "Low-res EELS" stands for the non-monochromated VG microscope.

As the course was aimed at a SEM-CL community, its content revolved around the competitive advantages of using STEM over SEM. To simplify, we insisted at the time on four points that we will more or less repeat in the presentation of this proceeding:

- The use of high-speed electrons, allowing to obtain a better spatial resolution (Figure 1, SI STEM-CL). The increase in spatial resolution was evident in the hyperspectral mapping of quantum confined systems [5]. For example, we had shown the possibility of spatially distinguishing two GaN quantum wells separated by a few nanometres in AlN wires. These types of studies, which are still flourishing (see for example [6]), will not be discussed in this paper.

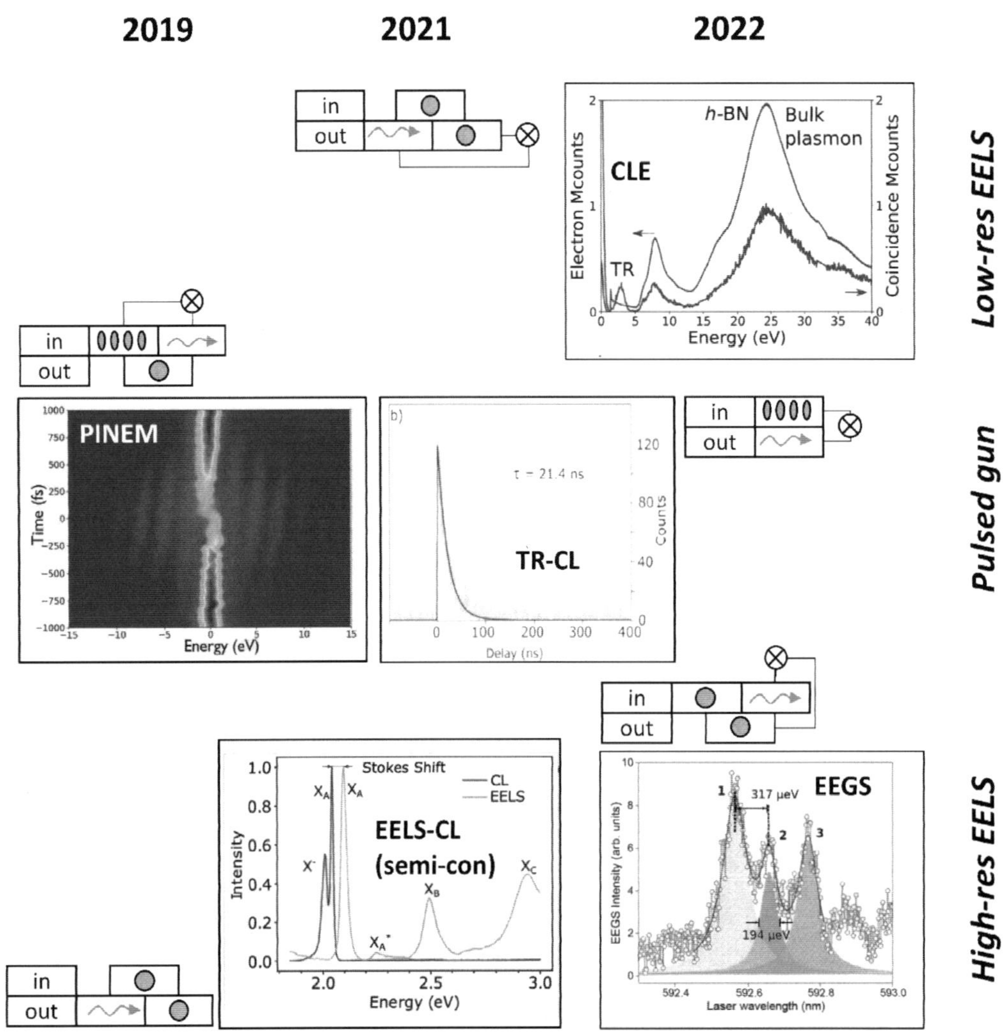

Figure 2. Some milestones experiments (from top to bottom, left to right [12, 11, 13, 14, 9]) after the lecture on STEM-CL at the 2015 edition of EMAS. They have been obtained on a continuous and non-monochromated VG microscope, a pulsed gun, non-monochromated Hitachi microscope (FEMTOTEM) and a highly monochromated NION microscope (CHROMATEM). PINEM: photon induced near-field electron microscopy; EEGS: electron energy gain spectroscopy; CLE: cathodoluminescence excitation spectroscopy; TR-CL: time resolved cathodoluminescence. The pictograms convention is given in Figure 1. "Pulsed gun" stands for the FEMTOTEM microscope and "High-res EELS" for the CHROMATEM microscope.

- The possibility of coupling CL to the vast range of STEM techniques. In particular, high resolution imaging had enabled a quantitative study of the link between emission wavelength and (atomic scale) quantum well size [5]. Perhaps most importantly, a comparative study of hyperspectral mapping of plasmons in energy loss spectroscopy (EELS) and CL had demonstrated the intimate link between optical (extinction/scattering) and electronic (EELS/CL) spectroscopies [4], see Figure 1. In this proceeding, we will see how the latest technological developments have made it possible to extend comparative EELS/CL analyses, which were previously limited to plasmons due to the poor spectral resolution of EELS, to optical excitations in semiconductors (Figure 2) or nanophotonic systems thanks to the emergence of monochromated machines. For the latter, we will see how the use of a CL detector to inject light leads to the creation of a new type of spectroscopy with very high spectral resolution (Figure 2) [7, 8, 9], the energy gain spectroscopy (EEGS), directly derived from photon induced near-field electron microscopy (PINEM) spectroscopies [10], techniques that are also possible with our mirror [11] (Figure 2).

- A gentle excitation of the material by high speed electrons, allowing to remain always in a linear excitation regime. This allowed us to demonstrate the similarity between STEM-CL and PL in the case of [15] quantum dots. Since then, this feature has been exploited for the study of single-photon emission through auto-correlation experiments using a flagship quantum optics system, the intensity interferometer (HBT) [3] (Figure 1). This was the first system of its kind for either STEM or SEM. The use of the HBT was unexpectedly successful not only in the STEM, but also later in the SEM, thanks to the discovery of an effect called "bunching" [16, 2, 17, 18], which allows a measurement of lifetimes at very high spatial resolution, and of which we will say a few words later. The time resolution limitations of HBT motivated the development of time-resolved cathodoluminescence in a TEM by the Toulouse team [13] (Figure 2). On the other hand, the power of photon-photon coincidence analysis led the Orsay team to develop also an EELS/CL coincidence analysis [12] (Figure 2), which will be discussed extensively in this proceeding.

2. Correlation between EELS and CL

In 2015, we showed that EELS and cathodoluminescence were the nanometric equivalents of extinction and diffusion spectroscopy for optics [4, 20]. Moreover, we had shown by comparing CL and PL on the same quantum dot that the two latter gave equivalent luminescence information [15]. It was, therefore, very tempting to extend the analogy valid for plasmons to semiconductors and photonic systems. Unfortunately, this required a much better spectral resolution in EELS than the few hundred meV accessible on an old generation microscope. Indeed, if the FWHM of plasmons in the visible is of the order of a few hundred meV and, therefore, compatible with old generation EELS technologies, the same is not true for excitons and photonic modes. The arrival of high-resolution monochromators [21] changed the situation in the mid-2010s, allowing, in addition to access to the infrared regime, to obtain resolutions below a few meV in the visible.

In Figure 3a we present the comparison between EELS and CL spectra for a monolayer of a transition metal dichalcogenide (TMD) two-dimensional semiconductor material (WS_2) [14]. These results are directly comparable to optical absorption and emission results, with the added spatial resolution of course. TMDs have recently become increasingly popular due to their spectacular optical properties. We can see directly the interest of using two different spectroscopies because they reveal different excitations within the same material. In particular, we have used them to study the formation of trions at the nanoscale [14].

It is also thanks to this generation of monochromators that it is now possible to study photonic structures in EELS [22], and thus to make relevant comparisons between EELS and CL for these systems. We show in Figure 3b a result obtained on a SiO_2-sphere of microns in diameter [9]. In this type of system it is known that there are modes propagating along the perimeter of the sphere and called WGM (whispering gallery modes). It is easy to see the interest of using both techniques: EELS allows access to a very wide range of modes, whereas CL has only a limited range. However, it is clear that if the signal to noise ratio (SNR) of EELS is much higher than that of CL, only the spectral resolution of the latter allows to resolve the modes (TE and TM).

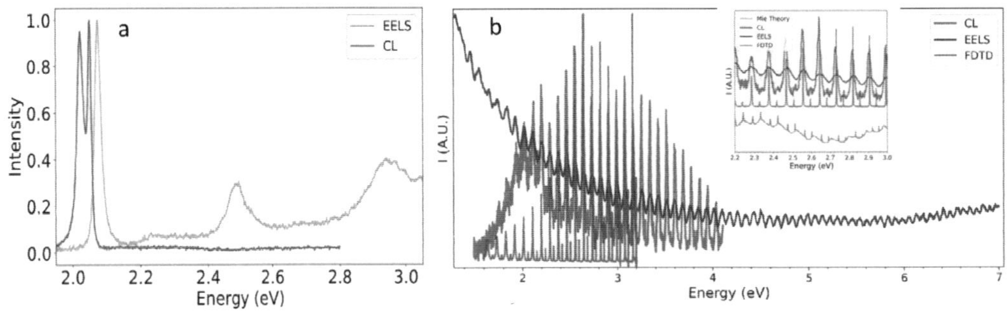

Figure 3. Comparison between EELS and cathodoluminescence (no time-coincidence). a) Typical electron energy loss spectrum (EELS, orange) and cathodoluminescence (CL, purple) at the same beam position on a WS_2 sample [14]. b) EELS and CL spectra of a µm-sized silica sphere measured by using 200 keV, compared to FDTD simulations [19]. Inset: Enlarged area showing the relatively low EELS resolution.

Despite the undeniable success of monochromation in addition to CL in a STEM, we can see that these techniques are limited in at least two ways. On the one hand, the spectral resolution of EELS and the SNR of CL are too low to expect to study photonic modes with high quality factors. On the other hand, the dynamics of the excitations cannot be studied.

The solution to these problems lies in the use of various synchronisation schemes between different signals as explained in figure 1 and 2.

3. Light injection in the (S)TEM

As predicted some 15 years ago, it is theoretically possible to accelerate the electrons by injecting a laser into the sample to pump the sample, creating an excitation that will accelerate the electrons. Such an energy gain is the stimulated counterpart of the energy loss [7]. This effect has been demonstrated using pulsed guns whose emission is synchronised with the illumination of a sample by a femtosecond laser [10]. Until very recently, the high temporal resolution of this type of set-up was not used and it was rather the strong interaction allowed by this type of illumination that was used. Our mirror could thus be used in FEMTOTEM to obtain a PINEM signal (see Figure 2) [11, 23]. The high numerical aperture requires a detailed simulation [23].

The use of a femtosecond laser, due to the energy-time uncertainty relation, does not allow spectral resolutions better than about 10 meV [24] to be achieved. Although remarkable considering the nominal resolution of the microscopes used, this is far from the necessary resolution (in the order of µeV!). The use of continuous or nanosecond lasers is therefore necessary [8, 9, 25]. The first demonstration of energy gain spectroscopy was made by the Ropers team using a continuous laser [8]. The exceptional µeV-resolution was however only obtained for a dedicated sample. We have recently used our mirror to obtain resolutions of the order of a hundred µeV in the case of an arbitrary dielectric sphere [9] (see Figure 2).

Now that we know that we can get rid of the EELS spectral resolution and CL SNR problems with the EEGS, can we study the dynamics of the excitations?

4. Lifetime measurements in the STEM

For this, the most obvious, but probably technology demanding, way is to use a pulsed gun to do cathodoluminescence in a STEM, as it is relatively common practice in the SEM [27]. This is what was done recently on the FEMTOTEM microscope in Toulouse where it was possible to map the lifetime of point defects in a nanodiamond with a spatial resolution of a few nanometres [13], as shown on Figure 4 (see also another STEM CL study at low voltage [28]). The lack of current and therefore of photons in STEM compared to SEM makes the experiment particularly complicated, and requires a mirror with a high numerical aperture in order to not lose too many photons.

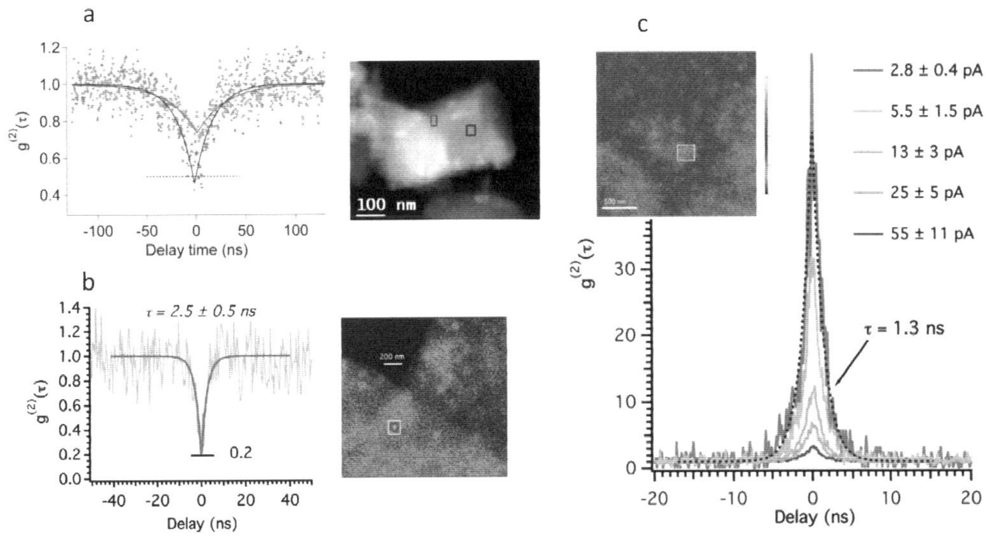

Figure 4. Intensity interferometry. a) $g^2(\tau)$ measurements on a nitrogen-vacancy centre in diamond, taken at two positions indicated on the left in the high angular dark field image [3]. b) Same for a defect in an h-BN flake, with the CL emission map on the right [26]. c) Same for a collection of the same defects [16]. In a) and b) the $g^2(\tau)$ displays antibunching, while in c) it displays bunching.

Temporal resolution and dynamic tracking can also be achieved with coincidence methods. We had indeed shown that it was possible to measure the autocorrelation function $g^2(\tau)$ as a function of the delay τ of a cathodoluminescence signal. The measurement of the autocorrelation function is a well-known method in optics to determine the quantum nature, or not, of a radiation. Thus, a value of the autocorrelation function smaller than 1 is the unequivocal signature of a purely quantum effect, in this case the emission of single photons. This effect has been measured by us in the case of single defects in nanodiamonds [3] and hexagonal boron nitride (h-BN) [26]. We also discovered that except for this particular case, the autocorrelation function was in general very large at zero delay [16]. This is one of the rare occasions when the PL and the CL are different. The apparent grouping of photons in CL comes from the fact that in CL it is essentially a volume plasmon that is excited, at high energy, and which de-excites very quickly in the form of several pairs of electron holes. The latter can all be de-excited at the same time, producing the synchronised emission of photons. "all at the same time" means "during the typical de-excitation time". It is, therefore, possible to measure the lifetime of excitations in systems, with ultimately the spatial resolution of STEM-CL [2]. It should be noted that the bunching measurement has been unequivocally successful for the measurement of lifetime [2, 17, 18] and quantum efficiency [17, 18], and especially in SEM. In this case, it is indeed a less effective alternative than the TR-SEM-CL, but much less restrictive experimentally because it does not require a pulsed gun or a fast blanker. Also, it permits to study directly the transition from single photon emission (anti-bunching) to bunching [29].

5. Cathodoluminescence excitation spectroscopy

As shown in Figures 1 and 2, beyond the correlation between signals, the synchronisation and coincidence between them clearly provides unique information. In particular, if the gain measurements (synchronisation between laser and electron) make it possible to obtain a very high spectral resolution, the TR-CL or the bunching measurement (electron-photon or photon-photon correlation) a high temporal resolution, there are still shady areas concerning the dynamics of excitations in materials. Let us take the case of Figure 3a. We know that the emission of photons (from a trion or exciton) follows an absorption of energy from the incident photon by the material. This absorption is necessarily contained in the EELS spectrum. But we have no way of knowing from which part of the EELS spectrum this energy comes: is it the X_A exciton?, X_B?, etc. In other words, we have no way of knowing how to trace the quantum efficiency of the material. Such a measurement is possible with photoluminescence (see Figure 5). In this technique (photoluminescence excitation spectroscopy, PLE), a monochromatic laser is used to excite the sample to a well-known energy. After measuring a PLE spectrum, it is then easy to relate the emission process to the absorption process. In the case of CL, the problem is that the electromagnetic field that follows the electron and is responsible for the absorption by the material is essentially white [30] (see Figure 5). Therefore, it is not possible to establish this link in general. However, in the case where an EELS event can be isolated in time, the situation changes. It is then possible to correlate the detection of an EELS event with that of the emission of a photon, and thus obtain a CLE spectrum [12]. This is shown in the spectrum of Figure 6a in the case of a hexagonal boron nitride (h-BN) flake. In the case of h-BN, the EELS spectrum is very different from the cathodoluminescence spectrum. The former is dominated by the absorption band and especially the bulk plasmon, while the latter is essentially dominated by the emission from a defect at 4.1 eV [26]. The CLE spectrum shows us which part of the absorption band creates the most photons. It is indeed the plasmon, as mentioned above. However, by taking the ratio of the CLE and EELS spectra, we can trace the energy-resolved quantum efficiency (Figure 5c). The intrinsic efficiency of an excitation to create luminescence is then determined, independently of its ability to absorb energy.

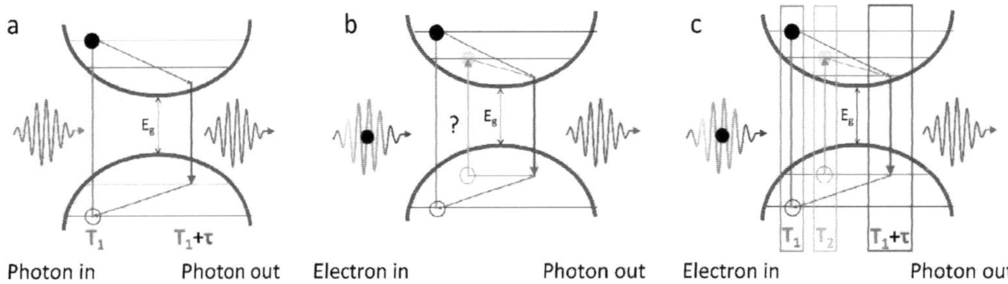

Figure 5. PLE and CLE. a) In PLE, a photon of defined wavelength (green) is absorbed at time T_1. After decay, a photon (red) may be emitted at time $T_{1+\tau}$. The path between the excited and de-excited state is easy to follow due to the monochromaticity of the incoming laser beam. b) In a regular EELS/CL experiment, an electron impinges on the sample. A (red) photon is emitted. The electromagnetic field following the electron is broad band, therefore, a set of states can be excited (the green and orange transitions), so it is impossible to follow the path between energy absorption and emission. c) In a CLE experiment, EELS and CL events are time-stamped, linking the photon emission to a specific EELS event at time T ($T = T_1$ in the present example).

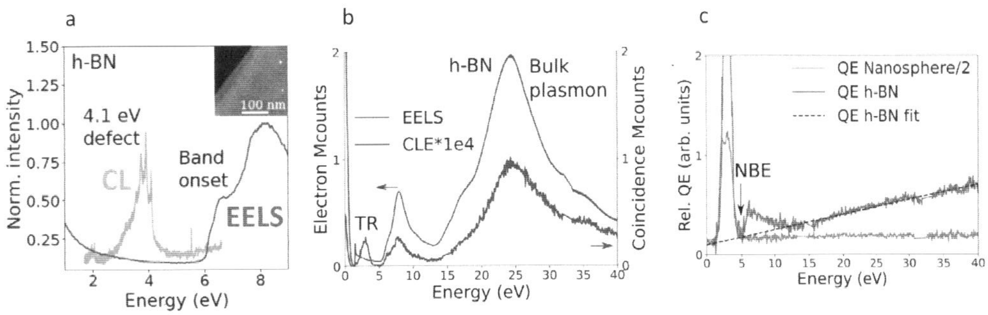

Figure 6. a) CL and EELS spectra of a thin h-BN flake. b) EELS and CLE spectra of another h-BN thin flake. c) Relative quantum efficiency (QE) of a plasmonic nanosphere and the thin h-BN flake of b). Near band edge (NBE) losses are more efficient pathways for light emission than energies below the bulk plasmon. Above 15 eV, the relative QE increases linearly. TR: transition radiations. From [12].

6. Conclusion and future prospects

We have seen that the use of a high numerical aperture mirror in a STEM is of interest for CL. As soon as the CL-signal is correlated with other signals, in particular the EELS, a whole range of new information emerges. The same is true when the mirror is used to inject light. We believe that the various techniques listed are still in their infancy, and that there is much to look forward to. Missing from the toolbox of our STEM-CL system, however, is the ability to polarise the light. It is important because, although phase-shaping with EELS holds great promises [31-33], polarisation cannot be accessed by EELS at the moment. On the opposite, this is routinely done in SEM [34], and in some TEM [35] or STEM [36]. In all these cases, the mirrors are either of small numerical aperture and/or of large size, which is not our case. However, we have recently succeeded in installing a cathodoluminescence polarisation system (see Figure 7), which makes us optimistic about the possibility of generalising the experiments presented above.

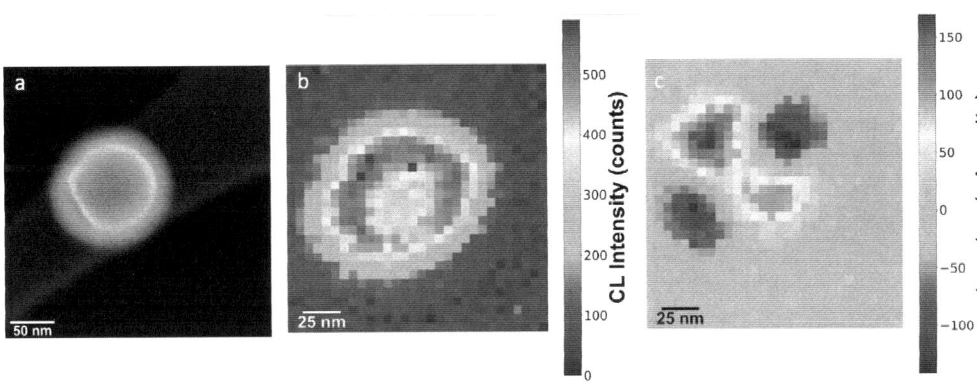

Figure 7. a) High angle annular dark field of a plasmonic sphere covered with a thin shell of silica. b) Non-polarised line. c) Linearly polarised CL maps filtered at the wavelength of the dipolar mode. In c), two orthogonal directions have been chosen.

References

[1] Kociak M and Zagonel L F 2017 *Ultramicroscopy* **174** 50-69
[2] Meuret S, Tizei L H, Auzelle T, Songmuang R, Daudin B, Gayral B and Kociak M 2016 *ACS Photonics* **3** 1157-1163
[3] Tizei L H and Kociak M 2013 *Phys. Rev. Lett.* **110** 1-5
[4] Losquin A, Zagonel L F, Myroshnychenko V, Rodriguez-Gonzalez B, Tence M, Scarabelli L, Forstner J, Liz-Marzan L M, de Abajo F J G, Stephan O and Kociak M 2015 *Nano Letters* **15** 1229-1237
[5] Zagonel L F, Mazzucco S, Tencé, March K, Bernard R, Laslier B, Jacopin G, Tchernycheva M, Rigutti L, Julien F H, Songmuang R and Kociak M 2011 *Nano Letters* **11** 568-573
[6] Hou J, Chen P, Shukla A, Krajnc A, Wang T, Li X, Doasa R, Tizei L H, Chan B, Johnstone D N, Lin R, Schülli T U, Martens I, Appadoo D, Ari M S, Wang Z, Wei T, Lo S C, Lu M, Li S, Namdas E B, Mali G, Cheetham A K, Collins S M, Chen V, Wang L and Bennett T D 2021 *Science* **374** 621-625
[7] García de Abajo F J and Kociak M 2008 *New J. Phys.* **10** 073035
[8] Henke J W, Raja A S, Feist A, Huang G, Arend G, Yang Y, Kappert F J, Wang R N, Möller M, Pan J, Liu J, Kfir O, Ropers C and Kippenberg T J 2021 *Nature* **600** 653-658
[9] Auad Y, Dias E J C, Tencé M, Blazit J D, Li X, Zagonel L F, Stéephan O, Tizei L H G, García de Abajo F J and Kociak M 2023 *Nature Comm.* **14** 4442
[10] Barwick B, Flannigan D J and Zewail A H 2009 *Nature* **462** 902-906
[11] Caruso G M, Houdellier F, Weber S, Kociak M and Arbouet A 2019 *Adv. Phys. X* **4** 166214
[12] Varkentina N, Auad Y, Woo S Y, Zobelli A, Bocher L, Blazit J D, Li X, Tencé M, Watanabe K, Taniguchi T, Stéphan O, Kociak M and Tizei L H G 2022 *Sci. Adv.* **8** eabq4947
[13] Meuret S, Tizei L H, Houdellier F, Weber S, Auad Y, Tencé M, Chang H C, Kociak M and Arbouet A 2021 *Appl. Phys. Lett.* **119** 062106
[14] Bonnet N, Lee H Y, Shao F, Woo S Y, Blazit J D, Watanabe K, Taniguchi T, Zobelli A, Stéphan O, Kociak M, Gradečak S and Tizei L H G 2021 *Nano Letters* **21** 10178-10185
[15] Mahfoud Z, Dijksman A T, Javaux C, Bassoul P, Baudrion A L, Plain J, Dubertret B and Kociak M 2013 *J. Phys. Chem. Lett.* **4** 4090-4094
[16] Meuret S, Tizei L H, Cazimajou T, Bourrellier R, Chang H C, Treussart F and Kociak M 2015 *Phys. Rev. Lett.* **114** 1-5
[17] Meuret S, Coenen T, Woo S Y, Ra Y H, Mi Z and Polman A 2018 *Nano Letters* **18** 2288-2293
[18] Finot S, Le Maoult C, Gheeraert E, Vaufrey D and Jacopin G 2022 *ACS Photonics* **9** 173-178
[19] Auad Y, Hamon C, Tencé M, Lourenço-Martins H, Mkhitaryan V, Stéphan O, García de Abajo F J, Tizei L H G and Kociak M 2022 *Nano Letters* **22** 319-327
[20] Losquin A and Kociak M 2015 *ACS Photonics* **2** 1619-1627
[21] Krivanek O L, Lovejoy T C, Dellby N, Aoki T, Carpenter R W, Rez P, Soignard E, Zhu J, Batson P E, Lagos M J, Egerton R F and Crozier P A 2014 *Nature* **514** 209-212
[22] Le Thomas N, Alexander D T, Cantoni M, Sigle W, Houdré R and Hébert C 2013 *Phys. Rev. B Cond. Matter Mater. Phys.* **87** 1-7
[23] Meuret S, Lourenço-Martins H, Weber S, Houdellier F and Arbouet A 2023 *ACS Photonics* **11** 977-984
[24] Wang K, Dahan R, Shentcis M, Kauffmann Y, Ben Hayun A, Reinhardt O, Tsesses S and Kaminer I 2020 *Nature* **582** 50-54
[25] Das P, Blazit J D, Tencé M, Zagonel L F, Auad Y, Lee Y H, Ling X Y, Losquin A, Colliex C, Stéphan O, García de Abajo F J and Kociak M 2019 *Ultramicroscopy* **203** 44-51
[26] Bourrellier R, Meuret S, Tararan A, Stéphan O, Kociak M, Tizei L H G and Zobelli A 2016 *Nano Letters* **16** 4317-4321
[27] Merano M, Sonderegger S, Crottini A, Collin S, Renucci P, Pelucchi E, Malko A, Baier M H, Kapon E, Deveaud B and Ganière J D 2005 *Nature* **438** 479-482

[28] Kim Y j and Kwon O h 2021 *ACS Nano* **15** 19480-19489
[29] Fiedler S, Morozov S, Komisar D, Ekimov E A, Kulikova L F, Davydov V A, Agafonov V N, Kumar S, Wolff C, Bozhevolnyi S I and Mortensen N A 2023 *Nanophotonics* **12** 2231-2237
[30] Polman A, Kociak M and García de Abajo F J 2019 *Nature Mater.* **18** 1158-1171
[31] Guzzinati G, Béché A, Lourenço-Martins H, Martin J, Kociak M and Verbeeck J 2017 *Nature Comm.* **8** 14999
[32] Lourenço-Martins H, Gérard D and Kociak M 2021 *Nature Phys.* **17** 598-603
[33] Lourenço-Martins H, Lubk A and Kociak M 2021 *SciPost Phys.* **10** 1-54
[34] Osorio C I, Coenen T, Brenny B J, Polman A and Koenderink A F 2016 *ACS Photonics* **3** 147-154
[35] Harvey T R, Henke J W, Kfir O, Lourenço-Martins H, Feist A, García De Abajo F J and Ropers C 2020 *Nano Letters* **20** 4377-4383
[36] Yamamoto N, Araya K and García de Abajo F J 2001 *Phys. Rev. B Cond. Matter Mater. Phys.* **64** 2054191-2054199

Interface bonding of co-laminated and annealed sheets – A short review of recrystallisation and grain growth mechanisms

F Brisset[1], H Azzeddine[2], D Bradai[3] and T Baudin[1]

[1] Université Paris-Saclay, CNRS, Institut de chimie moléculaire et des matériaux d'Orsay, 91400 Orsay, France
[2] Mohamed Boudiaf University, Faculty of Sciences, Laboratory of Materials and Renewable Energy, 28000 M'sila, Algeria
[3] University of Sciences and Technology - Houari Boumediene (U.S.T.H.B.), Faculty of Physics, Laboratory of Materials Physics, P.O. Box 32, El- Alia, Bab-Ezzouar, 16111 Algiers, Algeria

E-mail: francois.brisset@universite-paris-saclay.fr

Abstract. The roll bonding and the accumulative roll bonding allow the joining of two or more sheets of similar or dissimilar materials at various temperatures. This kind of metal-forming allows to access optimised physical properties of metallic composites. However, from a general point of view, the bonding strength of the co-laminated layers is very low and consolidation annealing is necessary to improve it. Consequently, recrystallisation and grain growth phenomena occur in the different layers and at the interfaces as well. The present short review highlights the recrystallisation and grain growth mechanisms such as continuous, discontinuous recrystallisation, particle-stimulated nucleation, and grain boundary migration mainly reported in the literature. Based on microstructural observations, it appears that consolidation annealing must be optimised to improve the bonding strength and the microstructural homogeneity of multilayer composites.

1. Introduction

Roll bonding (RB), which consists of co-laminating sheets composed of different metallic materials, is often used to combine the physical properties of each material [1]. Accumulative roll bonding (ARB) is derived from RB [2]. This metal-forming process was initially used as a severe plastic deformation process leading to an ultra-fine grain size giving in some cases superplastic properties at low temperatures [3, 4]. Then, it was employed to elaborate composites with new physical properties such as shielding ones combining Al and steel [5]. As illustrated in Figure 1, an ARB cycle consists of rolling two sheets together (for 50 % thickness reduction), cutting, wire brushing, and stacking them to achieve the initial thickness. The number of bonding interfaces, N_b, is calculated as follows: $N_b=2^N-1$, where N is the number of ARB cycles.

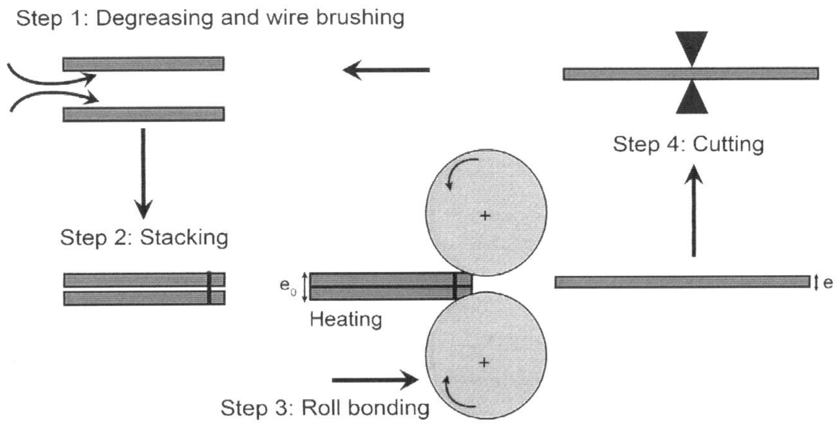

Figure 1. Schematic description of the ARB process principle: step 1: degreasing and wire brushing, step 2: stacking, step 3: roll bonding, and step 4: cutting.

The role of shear strain during the ARB process was studied by Lee *et al.* [6] and several papers were dedicated to the evolution of microstructure, texture and mechanical properties of metal, bi-metal and three-metal composites fabricated by the ARB process [5, 7-10]. For example, Tirsatine *et al.* [8] found that ARB processing of the Fe36%Ni (Invar) alloy up to 10 cycles lead to microstructural refinement with an average grain size of 0.5 μm. Boudekhani *et al.* [9] showed similar grain refinement (about 0.2 μm) in the case of Ni-14W (wt%) alloy ARB-processed up to 4 cycles. The tension tests of the ARB-processed Ni-14W alloy revealed a high strength (> 1,200 MPa) but very poor ductility [9]. The texture after ARB processing is mainly characterised by the typical rolling components (copper: {112}<111>, S: {123}<634>, and brass: {011}<211>) near the mid-thickness and shear texture (C: {001}<110>) at the surface of the sheets [8, 9]. In addition, Azzeddine *et al.* [11] reported the evolution of stored energy associated with the main texture components (copper, S, brass and cube: {001}<100>) as a function of the number of ARB cycles in Invar alloy by using neutron diffraction and kernel average misorientation approach, which plays a significant role in controlling recrystallisation nucleation. Recently, Habila *et al.* [10] demonstrated the formation of Mg17Al12 and Mg2Al3 phases near the interfaces bonding in Al1050/AZ31/Al1050 composite successfully fabricated by ARB at 400 °C up to 5 cycles due to elemental inter-diffusion between Al1050 and AZ31layers.

The mechanisms allowing the bonding of the composite layers are now quite known and they are based on the "fracture-extrusion-welding" model proposed by Bay [12]. During rolling, the work-hardened brushed surface breaks and the fresh metal of the opposite layers extrudes and welds between the hard fragments [12]. Several studies have been performed to improve layer adhesion by adjusting the sheet surface preparation, amount of thickness reduction especially for RB, rolling temperature, and strain rate [13].

Moreover, when the composite is fabricated of a hard layer included between two soft materials, the necking and the fragmentation of the central layer can appear facilitating the welding of the soft materials between the hard fragments [14]. So, in the case of an Al/steel/Al laminated sheet, Clérico *et al.* [1] have shown that the fragmentation of the steel layer allows the increase of the bonding strength starting from 60 % of thickness reduction during cold RB. In contrast, the fragmentation of steel in another Al/steel/Al composite fabricated by ARB processing appeared after 3 cycles and it was found to cause the reduction of the yield stress while improving the shielding [5].

However, after RB and ARB, the bonding strength of the co-laminated layers measured by peeling tests remains low [15]. Then, post-heat treatments are usually carried out to improve the interface welding by facilitating atomic diffusion and then activating the consolidation mechanisms [13]. Poncelet *et al.* [16] have calculated the theoretical diffusion distance of Ni element as a function of annealing temperature in bimetal composed of Fe-47.5Ni and Fe-50Ni-8.6Cr (wt.%) alloys. First, the distance was very short at 450 °C during 90 minutes, but it increases remarkably until around 12 µm at 1,200 °C during the same time which was in good agreement with the experimental value (16 µm) obtained by energy-dispersive X-ray spectrometry (EDS) measurement. Moreover, the annealing treatments provoke recovery, recrystallisation, and grain growth in different layers and at the bonding interfaces of the fabricated composite. For example, Figure 2 shows the evolution of the bonding strength measured by the peeling test as a function of the annealing temperature for roll-bonded bimetal Fe-Ni [16]. In addition, the evolutions of the recrystallisation fraction of both alloys as a function of annealing temperature are shown in Figure 2.

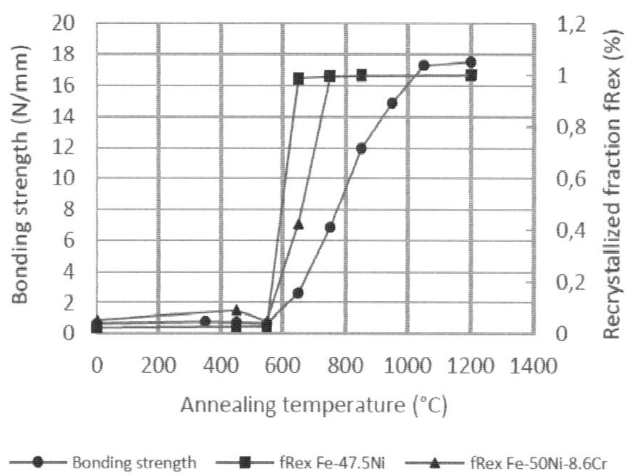

Figure 2. Effect of the annealing temperature (during 90 minutes) on the bonding strength [16].

As can be observed the bonding strength remains unchanged during annealing temperature below 600 °C. Then, the bonding strength increases significantly with increasing annealing temperature to reach a maximum value at 1,050 °C. Above 1,050 °C, a plateau is observed. The authors assume that this plateau is related to the formation of porosities due to the Kirkendall effect [17].

Figure 2 shows that the recrystallisation fraction (f Rex) in both alloys occurs at 550 °C and it is completed at 650 °C and 750 °C for Fe-47.5Ni and Fe-50Ni-8.6Cr alloys, respectively. It is interesting to note that even if the primary recrystallisation is complete in both alloys, the optimal bonding strength is not attained. Consequently, it can be concluded that the maximal value obtained after annealing treatment at 1,050 °C is due to the grain growth [16].

The evolution of recrystallisation and grain growth at the layers and near the interfaces of different multilayer composites are presented in the following sections.

2. Recrystallisation of layers

It is well known that due to the nature of ARB processing (see Figure 1), the microstructure of similar processed materials is heterogeneously distributed through the sample thickness [18, 19]. Hence, it is expected that the corresponding microstructural evolution kinetics will be more complicated in the case of dissimilar multilayer composite materials. Indeed, taking into account the stacking fault energy and the energy stored during the deformation processing, several kinds of primary recrystallisation mechanisms could occur during annealing treatments such as continuous or discontinuous recrystallisation, and particle stimulated nucleation (PSN) [20]. A continuous recrystallisation mechanism is involved when the sub-grains formed during deformation processing gradually migrated during annealing. While discontinuous recrystallisation mechanism occurs when recrystallisation proceeds by nucleation and growth of new grains within the deformed microstructure. Moreover, the PSN mechanism is considered as an important mechanism in multi-layered composite sheets due to the presence of different second-phase particles inside the bulk of co-laminated sheets, or oxide inclusions [20] and particles coming from the wire brushing process at the bonding interfaces [16, 22].

Duan et al. [20] have studied the microstructure and texture of Fe (bcc)/Ni (fcc)/Fe (bcc) multi-layered composite fabricated by ARB for four cycles followed by annealing treatments at 700 °C for 1, 5, and 30 min (Figure 3).

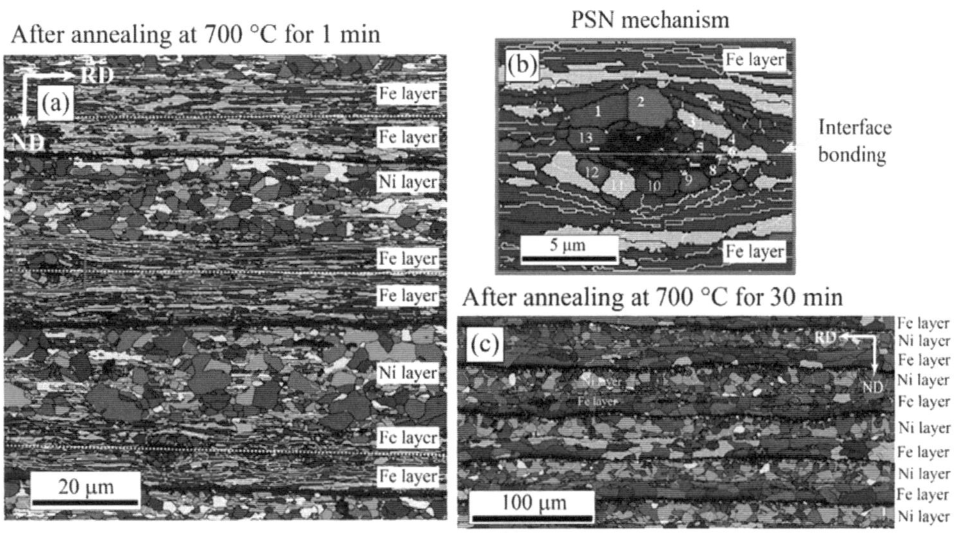

Figure 3. a & c) EBSD maps of ARB processed material after annealing for a) 1 min, and c) 30 min at 700 °C. b) High spatial resolution EBSD map showing PSN at the interface of Fe layers shown in a). The colours correspond to the different texture components [21]. Ni layers: yellow {011}<211>; blue {112}<111>; lime green {123}<634>; pink {011}<100> and red {001}<100>); Fe layers: purple α-fibre {hkl}<110> and light blue γ-fibre {111}<uvw>.

The results indicate that at the beginning of annealing, the Ni layers are partially recrystallised and continuous recrystallisation occurred, which maintains the rolling texture components, followed by discontinuous recrystallisation leading to the domination of the {100}<001> cube component. While the recrystallisation of the Fe layer begins at the Fe/Fe interfaces in the vicinity of oxide inclusion (Figures 3a and 3b) through the PSN mechanism leading to the formation of {001}<110> texture component and then a discontinuous recrystallisation mechanism dominates causing the formation of strong {511}<1 5 10> texture component after grain growth. The authors assumed that the combination of {100}<001> cube and {511}<1 5 10> components lead to a reduction in planar anisotropy and improve the formability of the Fe/Ni multilayered composite. Figure 3c shows the complete recrystallised microstructure after annealing for 30 minutes at 700 °C.

Other authors have observed that the recrystallisation microstructure at surface layers of similar ARB processed alloys such as pure Al and Ni-14%W alloy was characterised by larger grains and controlled by a continuous mechanism while a mixture of continuous and discontinuous mechanisms was observed in the interior layers resulting in small grain size [18, 23]. The results showed that the microstructural heterogeneity level decreases with an increasing number of ARB cycles [18].

3. Recrystallisation near the interface bonding

The interface bonding of multilayered composite is often characterised by EDS measurements to estimate the diffusion profile or to identify the intermetallic compounds formed at the interface of two materials [24, 25]. Kim *et al.* [24] have studied a roll bonded Al/Cu clad sheet (Figure 4a) after annealing at 400 °C during10, 20 (Figure 4b), 30, 60, 90 and 120 minutes (Figure 4c).

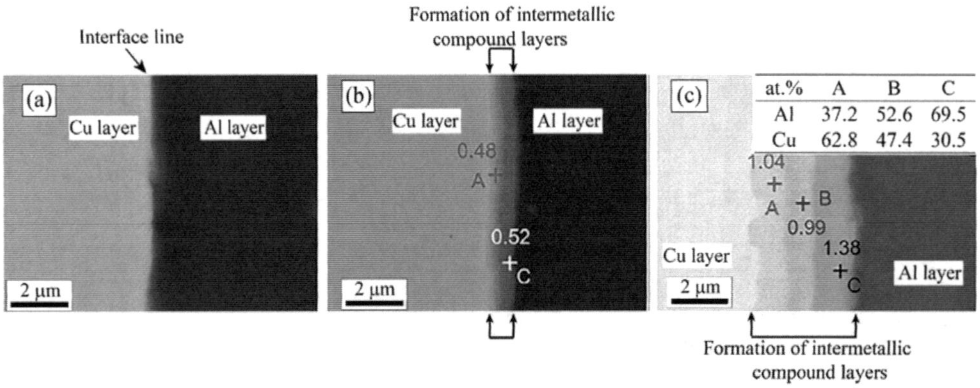

Figure 4. Scanning electron microscope (SEM) micrographs observed at the interface of a roll bonded Al/Cu clad after annealing at 400 °C during a) 0 (as-roll-bonded), b) 20 and c) 120 min. For 120 min, chemical composition (at%) is given for three different interface layers. The red, white, blue and dark values indicate the layer thickness [24].

After the annealing at 400 °C during 20 minutes, Figure 4b shows the appearance of two layers adjacent to copper and aluminium. After 30 minutes annealing and longer (Figure 4c, at 120 minutes), three layers are visible. The chemical composition indicated in Figure 4c shows the presence of three different intermetallic compound layers.

Nevertheless, only a few papers can be found in the literature dealing with the evolution of microstructure near the interface bonding using the electron backscatter diffraction (EBSD) technique [16, 18, 19, 21]. Some published results are described below.

As mentioned above, the PSN mechanism initiates the recrystallisation in the vicinity of oxide inclusions at the Fe/Fe interfaces [21]. Depending of the co-laminated material nature, intermetallic compounds can also be formed at the interfaces during the consolidation annealing and their distribution can affect strongly the bonding strength and the mechanical properties of the composite sheets [26, 27]. For example, the formation of a brittle Fe2Al5 intermetallic compound layer with 2 μm of thickness decreases the bonding strength of the tri-layered Al/brass/Al composite fabricated by the RB process [15]. While, the mechanical properties were enhanced due to the thickening of intermetallic compounds formed during the annealing of Al5052/Cu multilayer composites fabricated by ARB [27].

Work-hardened fragments coming from the wire brushing during the ARB preparation technique were found strongly affect the progress of recrystallisation [16, 22]. It is accepted that the fragments facilitate the recrystallisation of the matrix by creating high deformation zones as shown in Figures 5a and 5b for ARB processed Fe-47.5Ni/Fe-50Ni-8.6Cr (wt%) composite [16]. But the fragment recrystallisation kinetics is very low after annealing at 850 °C for 90 min compared to that of the matrix as demonstrated in Figure 5c which causes a delay for the layer bonding [16].

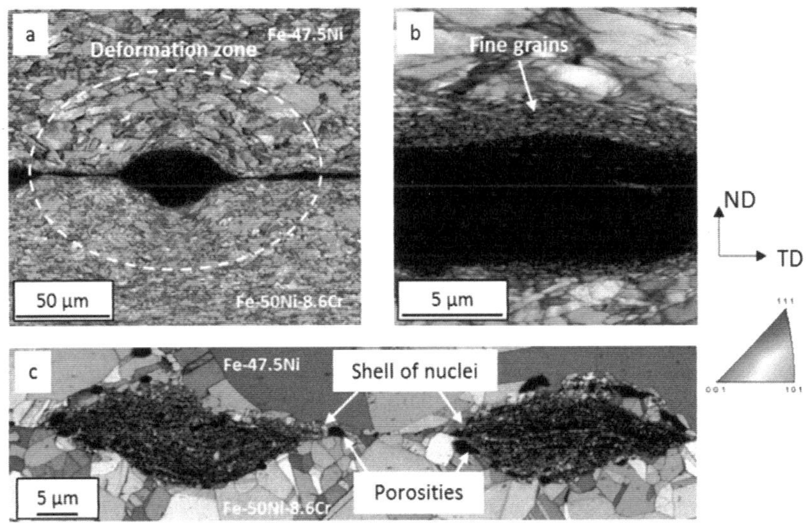

Figure 5. EBSD maps showing particles at the interface of two co-laminated Fe-Ni alloys: a) and b) before annealing, and c) after annealing at 850 °C for 90 min [16].

Long-time annealing or increasing annealing temperature is appeared necessary to suppress these fragments. In these conditions, the large grains in the matrix grow at the expense of the fragment small grains and finally grow through the sheet interfaces driven by grain boundary migration [28]. Some examples of grain growth in multilayer composites are shown in the next section.

4. Grain growth through the interfaces

Figure 6 shows the evolution of microstructure through the entire sample thickness of multi-layered copper sheets fabricated by ARB for 2, 4 and 6 cycles followed by subsequent annealing at 500 °C for 2h [29]. It must be noted that 5 copper layers were produced in the first rolling pass [29]. Then, the samples were composed of 10, 40 and 160 stacking layers after 2, 4 and 6 ARB cycles, as can be seen in Figures 6a, 6c and 6e, respectively. The authors note firstly that all the samples are completely recrystallised and secondly that obviously when the number of ARB cycles increases the grain size decreases since the dislocation density is higher and then the nucleation site density is also higher. The authors also remark that grain growth is limited by the thickness of the layers. This is true regarding the presence of porosities or work-hardened particles as discussed above. However, Figures 6b, 6d and 6e show clearly that during the grain growth of the samples some boundaries have crossed the layer interface (see the arrows) which guarantees successful bonding.

Similar interesting microstructural features were observed in ARB-processed Ni-14%W alloy and annealed at 1,100 °C for 1 h [18]. Figure 7a depicts a selected region from the EBSD measurements of the annealed sample showing the grains having sizes of layer zone (~ 50 µm).

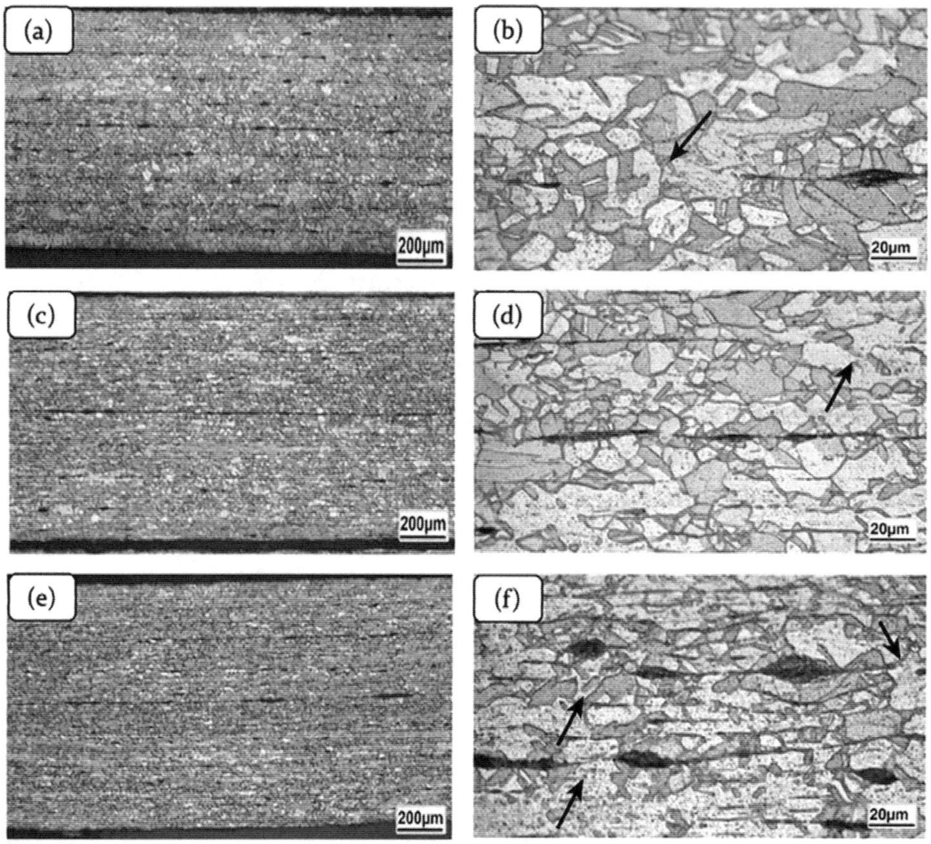

Figure 6. Optical micrographs of copper samples fabricated by ARB and annealed at 500 °C for 2h: a) and b) 2 cycles, c) and d) 4 cycles, and e) and f) 6 cycles [29].

Figure 7a shows another zone indicating the formation of small grains along the bonding interface due to the presence of particles resulting from the wire-brushing procedure. Let us remark that it was also found that the oxide inclusions tend to retard the grain boundary migration through the interface in aluminium 1100 alloy ARB processed for 8 cycles and annealed at 300 and 400 °C and it then retards grain growth. [30]. In fact, the heterogeneity of recrystallised grain size through the multilayered composite is also associated with the grain boundary network. The large grains were found surrounded by random grain boundaries which promote their migration [18]. While the small grains near the interfaces were found surrounded by inactive special CSL boundaries [18].

At the same time, Figure 7b shows also the migration of grain boundaries and twins across the bonding interface. It is believed that the bonded interfaces are eliminated with an increasing number of ARB cycles and transformed themselves to migrated interfacial grain boundaries during grain growth, which improves the bonding of the composite material [18]. Even if a grain boundary migrates, it leaves a "ghost" line at the original position of the bonded interface as can be observed in Figure 7b. This kind of ghost boundary has often been observed and corresponds to thermal grooves left by the initial annealing and it can be used for quantifying the grain boundary motion [31].

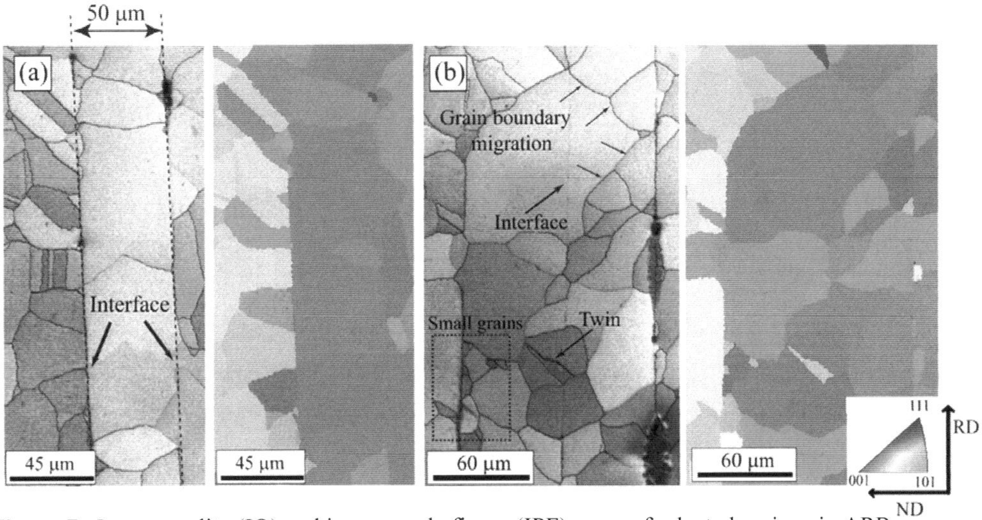

Figure 7. Image quality (IQ) and inverse pole figure (IPF) maps of selected regions in ARB-processed Ni-14%W alloy for 4 cycles and annealed at 1,100 °C for 1h showing: a) grains having sizes of layer zone, and b) migration of grain boundary, twin and small grains along the bonding interface [18].

From this qualitative analysis essentially based on microstructural observations, it is apparent that consolidation annealing is the best procedure to improve the bonding strength of multilayer composites owing to the enhancement of atomic diffusion and the occurrence of primary recrystallisation and grain growth. Nevertheless, the heterogeneous microstructure is still obtained at the layers and at the interfaces due to the nature of ARB processing, the presence of different second particles and grain boundary network. Much work is required to extend our knowledge of the mechanisms responsible for primary recrystallisation and grain growth at the layer interface, especially in the presence of hard fragments.

5. Conclusion

The present short review deals with the evolution of recrystallisation and grain growth microstructure near the interface bonding of multi-layered composites fabricated by roll bonding and accumulative roll bonding processing. The main conclusions can be summarised as follows:

- The bonding interface strength can be improved during deformation processing based on the "fracture-extrusion-welding" model. However, the bonding strength remains low and a consolidation annealing must be performed.
- Continuous and discontinuous recrystallisation mechanisms were detected inside the layers. While particle stimulated nucleation mechanism dominates near the interface bonding.
- Metal fragments, oxides, and intermetallic compounds generated during accumulative roll bonding retard the primary recrystallisation and constitute barriers for the reinforcement of bonding interface. Long-time annealing is necessary to erase these barriers.
- Grain growth allows grain boundary migration across the bonded interfaces which guarantees successful bonding.
- The grain growth is more effective in improving the interface bonding than primary recrystallisation.

IOP Conf. Series: Materials Science and Engineering **1324** (2025) 012004 doi:10.1088/1757-899X/1324/1/012004

References

[1] Clérico P, Mininger X, Prévond L, Baudin T and Helbert A L 2019 *J. Alloys Compounds* **798** 67-81

[2] Tsuji N, Saito Y, Lee S.H and Minamino Y 2003 *Adv. Eng. Mater.* **5** 338-344

[3] Jafarian H R, Mousavi Anijdan S H and Miyamoto H 2020 *Mater. Sci. Eng. A* **795** 140015

[4] Ebrahimi M and Wang Q 2022 *J. Mater. Res. Technol.* **19** 4381-4403

[5] Verstraete K, Prévond L, Helbert A L and Baudin T 2019 *Adv. Eng. Mater.* **21** 1800967

[6] Lee S H, Saito Y, Tsuji N, Utsunomiya H and Sakai T 2002 *Scripta Materialia* **46** 281-285

[7] Mahdavian M M, Ghalandari L and Reihanian M 2013 *Mater. Sci. Eng. A* **579** 99-107

[8] Tirsatine K, Azzeddine H, Baudin T, Helbert A L, Brisset F, Alili B and Bradai D 2014 *J. Alloys Compounds* **610** 352-360

[9] Boudekhani S, Azzeddine H, Tirsatine K, Baudin T, Helbert A L, Brisset F, Alili B and Bradai D 2018 *J. Mater. Eng. Performance* **27** 4561-4570

[10] Habila W, Azzeddine H, Mehdi B, Tirsatine K, Baudin T, Helbert A L, Brisset F, Gautrot S, Mathon M H and Bradai D 2019 *Mater. Charact.* **147** 242-252

[11] Azzeddine H, Tirsatine K, Baudin T, Mathon M H, Helbert A L, Brisset F and Bradai D 2017 *Mater. Chem. Phys.* **201** 408-415

[12] Bay N 1983 *Welding Res. Suppl.* **62**(5) 137-142

[13] Li L, Nagai K and Yin F 2008 *Sci. Technol. Adv. Mater.* **9** 023001

[14] Utsunomiya H, Maeda S, Imai T and Matsumoto R 2018 *CIRP Annals – Manuf. Technol.* **67** 317-320

[15] Naseri M, Reihanian M and Borhani E 2016 *J. Manuf. Process.* **24** 125-137

[16] Poncelet J, Baudin T, de Oliveira M, Waeckerlé T, Ateba-Betanda Y, Brisset F and Helbert A L 2019 *J. Physics* **1270** 012024

[17] Philibert J, Vignes A, Bréchet Y and Combrade P 2002 *Métallurgie - Du minerai au matériau* (2nd edition). [Paris, France: Dunod]

[18] Koriche S, Boudekhani-Abbas S, Azzeddine H, Baudin T, Helbert A L, Brisset F, Ateba Betanda Y, Waeckerlé T and Bradai D 2020 *SN App. Sci.* **2** 650

[19] Tirsatine K, Azzeddine H, Baudin T, Helbert A L, Brisset F and Bradai D 2017 *Mater. Eng.* **24** 56-66

[20] Humphreys F J and Haterly M 2004 *Recrystallization and related annealing phenomena* (2nd edition). [Oxford, UK: Elsevier]

[21] Duan J Q, Quadir M Z, Xu W, Kong C and Ferry M 2017 *Acta Materialia* **123** 11-23

[22] Mishin O, Zhang Y and Godfrey A 2017 *J. Mater. Sci.* **52** 2730-2745

[23] Li S, Yang L and Qin N 2018 *J. Mater. Sci. Technol.* **34** 821-831

[24] Kim M J, Lee K S, Han S H and Hong S I 2021 *Mater. Character.* **1745** 111021

[25] Mahdavian M M, Khodabandeh A R, Jafarian H R and Mirdamadi Sh 2022 *J. Alloys Compounds* **925** 166711

[26] Wang C, Liu, X Jiang Y, Xie J, Zhang X and Zhou D 2018 *Sci. Bull.* **63** 1448-1456

[27] Tayyebi M, Rahmatabadi D, Karimi A, Adhami M and Hashemi R 2021 *J. Alloys Compounds* **871** 159513

[28] Poncelet J 2023 *Placage d'alliages à propriétés fonctionnelles : microstructure, adhésion et nouveaux matériaux*. PhD thesis. [Orsay, France: Université Paris-Saclay]

[29] Eivani A R, Shojaei A, Salehi M T, Jafarian H R and Park N 2021 *J. Mater. Res. Technol.* **10** 291-305

[30] Kwan C, Wang Z and Kang S B 2008 *Mater. Sci. Eng. A* **480** 148-159

[31] Yin W, Wang W, Fang X, Qin C and Xing X 2015 *Mater. Character.* **107** 134-138

Subscale inversion of X-ray emission in electron probe microanalysis based on deterministic transport equations

T Claus[2], G Achuda[1], S Richter[1] and M Torrilhon[2]

[1] RWTH Aachen University, Central Facility for Electron Microscopy, Aachen, Germany
[2] RWTH Aachen University, Applied and Computational Mathematics, Aachen, Germany

E-mail: mt@acom.rwth-aachen.de

Abstract. The reconstruction methods (ZAF, $\phi(\rho z)$ matrix correction) commonly applied in electron probe microanalysis (EPMA) make restrictive assumptions about the material structure. The model assumption that the material is homogeneous (or layered in depth) limits the spatial resolution of the reconstruction method to the interaction volume of electrons. Models based on Monte Carlo (MC) methods allow for arbitrary material structure, but suffer from statistical noise that prevents the use of MC in gradient-based reconstruction. We present a reconstruction framework for EPMA based on deterministic transport equations to model electron transport. Similar to MC, our model allows for an arbitrary material structure, but is also well-suited for application in gradient-based reconstruction. In this paper we discuss our model with a focus on inversion and discuss the necessity of regularisation (parametrisation of the material) for the reconstruction method. We conclude with reconstruction examples in 1D and 2D, that exemplarily show different parametrizations and, more importantly, the opportunity to resolve material structure smaller than the interaction volume.

1. Introduction

In electron probe microanalysis (EPMA) the sample is excited by a focussed beam of electrons which generates X-rays that are characteristic to the chemical components of the sample. The measured X-ray intensity $I_{(Z,j)}$ for atomic number Z of the associated atom and a transition j is normalised into an experimental k-ratio $k^{exp}_{(Z,j)} = I_{(Z,j)}/I^{std}_{(Z,j)}$ by the intensity of a standard sample.

An experiment yields multiple k-ratios: one for each of the considered characteristic X-ray transitions. Additionally, the experimental setup U, e.g., the beam position, energy or angle, can be varied to obtain more information about the material. To reconstruct the spatial distribution of the chemical composition of the sample, the collection of all k-ratios $k = \left\{k^{U_1}_{(Z,j)}, \dots, k^{U_2}_{(Z,j)}\right\}$ must be post-processed by an inversion algorithm, see also Figure 1.

In this work, we consider spatially varying mass concentrations $\rho : \mathbb{R}^3 \to \mathbb{R}^{n_e}, x \mapsto \rho(x)$ as the general description of compounds in the sample. Thereby $x \in \mathbb{R}^3$ is the spatial coordinate, n_e is the number of constituents. The X-ray intensity will depend on $\rho(x)$, and, hence, the general reconstruction problem is given by

$$\rho^*(x) = \arg\min_{\rho(x)} \operatorname{disc}(k(\rho(x)), k^{\exp}). \tag{1}$$

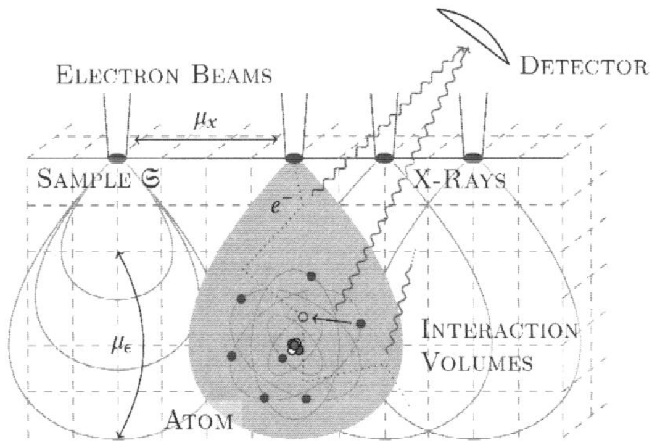

Figure 1. To increase the resolution of EPMA we use multiple experiments with electron beams at different positions μ_x and different energies μ_ϵ. Once the interaction volumes inside the sample start to overlap, a subscale inversion must use all experimental data to extract a spatial variation of mass concentrations.

That means, we search for the argument of the minimum (arg min), resp. the distribution of mass concentrations $\rho^*(x)$ that minimises the discrepancy function disc$(\cdot, \cdot)$ between modelled $k(\rho(x))$ and measured k^{exp} k-ratios. The unique reconstruction of the infinite dimensional field $\rho^*(x)$ from a finite set of measurements k^{exp} is only possible with additional constraints or prior assumptions. Additionally, the precise representation of uncertainty in $\rho^*(x)$ is a useful approach. In the mathematical framework of inverse problem theory, knowing the uncertainty of the unknown $\rho^*(x)$ is defined as the solution of the inverse reconstruction problem. Prior knowledge about the material and the additional knowledge based on the experimental measurements reduces general ignorance of $\rho^*(x)$.

1.1. Reconstructions and regularisation

The representation of the unknown $\rho(x)$ as a function has infinite dimensions. Therefore, it cannot be reconstructed from a finite set of data k^{exp} even in the absence of noise. The inverse problem is considered as being ill-posed, because no unique solution exists. To mitigate the ill-posed behaviour of the inverse problem, regularisation adds information. Thus, it is closely connected to the definition of prior knowledge about the material ρ from inverse problem theory.

Regularisation is a far-reaching strategy that can be implemented in many forms. Benning and Burger [1] gives a comprehensive overview over classical and modern regularisation techniques. Often, they rely on the assumption of a certain regularity of the parameters, concerning e.g., their value or their gradient. A well-studied regularisation technique is Tikhonov regularisation, that adds a penalty function based on the norm of the parameters to the minimisation problem.

In EPMA we mainly think of regularisation as the definition of a parametrisation of the material description using $\rho(x; p)$. Instead of specifying the material as a function $\rho(x)$ with infinite dimensions, we reduce the dimensionality to a finite dimensional parameter vector p, which parametrises the mass concentrations $\rho(x; p)$. The parametrisation translates the general reconstruction problem searching for mass concentrations $\rho^*(x)$, into the reconstruction problem searching for parameters p^* that describe structure in the material.

Reducing the variability of the unknown $\rho(x)$ into a predefined function $\rho(x; p)$ must be accompanied by the addition of carefully chosen material knowledge, because the specification of a parametrisation predefines all possible reconstruction results. Furthermore, the parametrisation should fit the available measurements. The more flexible the parametrisation is, the more measurements have to be provided, otherwise the measurements lack enough information and the reconstruction result is ambiguous.

IOP Conf. Series: Materials Science and Engineering **1324** (2025) 012005 doi:10.1088/1757-899X/1324/1/012005

1.2. A deterministic EPMA inversion framework

The diversity of questions which are posed to EPMA illustrates the need to develop a reconstruction method that remains general in the material parametrisation. We base our model on the same mathematical framework as Monte Carlo models. The electron transport is governed by the Boltzmann equation, however, instead of random sampling, we rely on a deterministic approximation of the Boltzmann equation [2-5]. This allows the exact computation of gradients, the main building block of reconstruction methods. Gradients based on Monte Carlo models are polluted by statistical noise and their use in reconstruction methods is restricted.

On top of the k-ratio model, we propose to use a combination of algorithmic differentiation [6, 7] and the adjoint state method [8] to compute gradients of the k-ratio model. Algorithmic differentiation is a method to compute gradients of numerical functions in a systematic (usually even automatic) and efficient manner. The adjoint state method describes a way to efficiently differentiate models based on partial differential equations (our Boltzmann approximation). The combination of both allows us to remain generic both in the objective function and the material parametrisation.

The implementation of the forward model as well as the inversion procedure is done in the programming language **julia**, which is a rapidly growing coding framework focussing on numerical methods and scientific computing. While providing a convenient programming syntax, it also enables high computational efficiency due to its just-in-time compiler and the concept of dynamic dispatch. It might also be the language of choice for further development of the reconstruction method, because using the concept of dynamic dispatch, various optimisations of the reconstruction method based on different parametrisations can be implemented conveniently.

2. Deterministic forward modelling

We use a deterministic k-ratio model pioneered in [4, 5]. The material is specified by its distributions of mass concentrations as the input to the forward problem $\rho(x)$. Below we discuss the successive parts in the computation of k-ratios.

2.1. The intensity of generated X-rays

The intensity of generated X-rays $I_{(Z,j)}$ is proportional to a composition of multiple consecutive model parts (the symbol $\circ$ means function composition, e.g., $(f \circ g)(x) = f(g(x))$):

$$I_{(Z,j)}(\rho) \propto \mathcal{A}_{(Z,j)}(\rho) \circ \mathcal{X}_{(Z,j)}(\rho) \circ \phi_{(Z,j)} \circ \Psi(\rho). \tag{2}$$

When comparing to experimental k-ratios, the proportionality constant cancels due to the normalisation into k-ratios. Beginning with the innermost model part, they are given by the following processes. The transport of induced electrons inside the sample, quantified by the electron fluence $\Psi(\epsilon, x, \Omega) = |v(\epsilon)| n(\epsilon, x, \Omega)$, where $n(\epsilon, x, \Omega)$ is the number density of electrons with energy ϵ at position x moving in direction Ω and $v(\epsilon)$ the velocity of electrons. The ionisation distribution $\phi(x)$ of atoms leading to the emission of X-rays. The X-ray generation distribution $\mathcal{X}(x)$ which combines the ionisation distribution $\phi(x)$ with the actual presence of atoms, and the attenuation $\mathcal{A}$ experienced by the X-rays due to absorption and scattering.

While in Eq. (2) direct dependence of the model operators on the mass concentrations is highlighted by $\cdot\,(\rho)$, later we highlight the dependency of a realisation of each operator on the electron energy ϵ, the position x and the direction Ω. Note that through the parts of the model, the dimensionality of the variables decreases from $(\epsilon, x, \Omega) \in \mathbb{R}^+ \times \mathbb{R}^3 \times S^2$ to a scalar $I_{(Z,j)} \in \mathbb{R}$.

X-rays, which are generated inside the sample, are attenuated while passing through to reach the detector. The attenuation of X-rays which are travelling along an axis (described by $z \in \mathbb{R}$) is commonly modelled using the Beer-Lambert law [9], which in the case of a constant attenuation coefficient μ simplifies to

$$\mathcal{I}(z) = \mathcal{I}_0 \exp(-\mu z). \tag{3}$$

IOP Conf. Series: Materials Science and Engineering **1324** (2025) 012005 doi:10.1088/1757-899X/1324/1/012005

In our model, X-rays are generated from a continuous field, hence the initial intensity is $I_0 = \mathcal{X}(x)$. The attenuation factor for a position x in the material is the line integral from x to the detector $\int_{d(x)} \cdot \, d\bar{z}$. The detected intensity is then given by the integral over the whole material.

$$\mathcal{A}_{(Z,j)}(\mathcal{X}_{(Z,j)}) = \int_{\mathbb{R}^3} \exp\left(-\int_{d(x)} \mu_{(Z,j)}(\bar{z})\,\mathrm{d}\bar{z}\right) \mathcal{X}_{(Z,j)}(x)\,\mathrm{d}x. \tag{4}$$

The linear attenuation coefficient $\mu_{(Z,j)}(x)$ for compounds can be modelled by linear combination with mass concentrations ρ_i in the form $\mu_{(Z,j)}(\chi) = \sum_{i=1}^{n_e} \rho_i(\chi) \left(\frac{\mu}{\rho}\right)_{i,(Z,j)}$ whereby the mass attenuation coefficients $\left(\frac{\mu}{\rho}\right)_{i,(Z,j)}$ for elements i are interpolated for the characteristic energy of an X-ray (Z,j). Implementations and data for mass absorption coefficients can be found in [10-12].

To compute the X-ray generation distribution for an X-ray (Z,j), we weigh the ionisation distribution $\phi_{(Z,j)}$ by the number of atoms of element Z per unit volume $\frac{\rho_Z}{A_Z}$ and the fluorescence yield $\omega_{(Z,j)}$ for the transition (Z,j)

$$\mathcal{X}_{(Z,j)}(\phi_{(Z,j)}) = \frac{\rho_Z(x)}{A_Z}\omega_{(Z,j)}\phi_{(Z,j)}(x). \tag{5}$$

Thereby $\rho_Z(x)$ is the mass concentration and A_Z the absolute atomic mass of element Z.

The ionisation distribution field $\phi_{(Z,j)}(x)$ of X-ray (Z,j) is given by product of the number density of electrons $n(\epsilon, x, \Omega)$ travelling with velocity $v(\epsilon)$ in direction Ω and the ionisation cross-section $\sigma_{(Z,j)}(\epsilon)$ describing the fraction of collisions leading to core shell ionisations which may result in the emission of (Z,j) X-rays

$$\phi_{(Z,j)}(x) = \int_0^\infty \sigma_{(Z,j)}(\epsilon) \int_{S^2} \underbrace{|v(\epsilon)| n(\epsilon, x, \Omega)}_{=:\Psi(\epsilon,x,\Omega)} \mathrm{d}\Omega\,\mathrm{d}\epsilon \;. \tag{6}$$

Implementations and data for ionisation cross-sections and fluorescence yields can be found in [11, 13-17].

Due to the high dimensionality ($\mathbb{R}^+ \times \mathbb{R}^3 \times S^2$) of the electron number density $n(\epsilon, x, \Omega)$ and the electron fluence $\Psi(\epsilon, x, \Omega)$, a simulation of either is complex. Hence, it is advantageous to derive a model for the averaged electron fluence. In Eq. (6), the electron number density only appears weighted by the magnitude of velocity $|v(\epsilon)|$ and averaged in direction $\Omega \in S^2$. We define

$$\psi_0^0(\epsilon, x) = \int_{S^2} |v(\epsilon)| n(\epsilon, x, \Omega)\,\mathrm{d}\Omega \tag{7}$$

and describe the P_N model, a moment expansion that governs ψ_0^0 (the zeroth moment of the electron fluence Ψ), in the next section.

2.2. *The electron fluence: The P_N-moment-model*

For a detailed derivation of the P_N model see [18, 19]. Here only the expansion using the method of moments [2] is described.

The linear Boltzmann equation [20] describes particle transport, where the self-interaction between particles can be neglected and only (elastic and inelastic) scattering with the background medium is considered. In the context of EPMA it is commonly assumed that the time dependence of the linear Boltzmann equation is negligible, because during the finite time interval of electron exposure, the stationary distribution is reached almost instantly. Typically, also the inelastic scattering operator is replaced by its continuous slowing down approximation (CSD). The continuous energy loss is modelled using the stopping power $S(\epsilon, x)$, the average energy loss per path length. The stopping power governs parts of the scattering operator in the linear Boltzmann equation, the leftover parts we call $Q(\epsilon, x)[\Psi]$. The resulting transport equation is an evolution equation for the electron fluence $\Psi(\epsilon, x, \Omega) = |v(\epsilon)| n(\epsilon, x, \Omega)$ given by:

$$-\partial_\epsilon (S(\epsilon, x)\Psi(\epsilon, x, \Omega)) + \Omega \cdot \nabla_x \Psi(\epsilon, x, \Omega) = Q(\epsilon, x)[\Psi(\epsilon, x, \Omega)]. \tag{8}$$

2.2.1. Spectral Galerkin approximation. As mentioned, it is expensive to solve Eq. (8) numerically, hence we will reduce the model complexity using a Galerkin method, which expands $\Psi(\epsilon, x, \Omega)$ into a linear combination of moments and basis function in Ω. The expansion is called the method of moments [2]. The spherical harmonics [21], an orthonormal set of functions $S^2 \to \mathbb{R}$, have proven to be suitable basis functions, because they possess favourable properties, which are inherited to the resulting model (the P_N model). The real spherical harmonic $Y_l^k: S^2 \to \mathbb{R}$ of degree $l \in \mathbb{N}^0$ and of order k ($|k| \leq l$) maps a direction $\Omega(\mu, \varphi) \in S^2$, also represented by polar $\mu \in [0, \pi]$ and azimuthal (longitudinal) angle $\varphi \in [0, 2\pi]$, to the real numbers $\mathbb{R}$.

We apply the method of moments to the BCSD as given in Eq. (8). The electron fluence Ψ is approximated by a linear combination of spherical harmonics up to degree $l \leq N \in \mathbb{N}^0$

$$\Psi(\epsilon, x, \Omega) \approx \Psi_{P_N}(\epsilon, x, \Omega) = \sum_{l \leq N, |k| \leq l} \psi_l^k(\epsilon, x) Y_l^k(\Omega). \tag{9}$$

The coefficients $\psi_l^k(\epsilon, x)$ are the moments of the electron fluence Ψ and, since spherical harmonics form an orthonormal basis of S^2, the moment ψ_l^k is given by the scalar product

$$\psi_l^k(\epsilon, x) = \int_{S^2} \Psi(\epsilon, x, \Omega) Y_l^k(\Omega)\, \mathrm{d}\Omega. \tag{10}$$

Moments ψ_l^k of higher degree $l > N$ are assumed to be $\ll 1$, therefore negligible. The zeroth spherical harmonic $Y_0^0(\Omega)$ is constant, so we can identify $\psi_0^0(\epsilon, x)$ with the variable in Eq. (6).

From Eq. (8) evolution equations for the moments ψ_l^k are derived by testing (multiplying and integrating: $\int_{S^2} Y_{l'}^{k'}(\Omega) d\Omega$ with spherical harmonics. Testing with the set of same spherical harmonics ($l' \leq N$, $|k'| \leq l'$), which are used in the approximation Ψ_{P_N} leads to a system of equations, the P_N model

$$-\partial_\epsilon (S\psi) + A^{(x)} \partial_x \psi + A^{(y)} \partial_y \psi + A^{(z)} \partial_z \psi + C\psi = 0. \tag{11}$$

The unknown variables are the coefficients $\psi = \{\psi_l^k\}_{l \in \mathbb{N}_0 |k| \leq l}$ used in the P_N approximation, the moments of the electron fluence Ψ. We derive the summands in Eq. (11) from Eq. (8) individually.

2.2.2. Stopping power. The first term in Eq. (8) is linear in Ψ and the spherical harmonics form an orthonormal basis, hence the term diagonalises in the moment method. The stopping power S is a material property, where for a compound we derive additivity using the mass concentrations $\rho(x)$ from the additivity approximation for cross-sections [22]

$$S(\epsilon, x) = \sum_{i=1}^{n_e} \rho_i(x) \frac{S_i(\epsilon)}{A_{Z_i}}. \tag{12}$$

Thereby $S_i(\epsilon)$ is the specific stopping power of pure element i. For the stopping power of pure elements there exists multiple models, e.g., a model based on the Bethe-Loss formula [23] or interpolation tables from a more advanced model [16, 24]; we use the latter in our current implementation.

2.2.3. Transport coefficient. The transport coefficient C is derived from the scattering operator $Q(\epsilon, x)[\Psi]$ that models elastic and deflections of inelastic collisions (the energy loss is modelled by the stopping power). The scattering operator Q is linear in the electron fluence Ψ, so if we replace Ψ with its P_N approximation and test with the spherical harmonic $Y_{l'}^{k'}$, we derive the transport coefficient matrix $\mathcal{C}$ by similar argument as for the stopping power

$$C_{(l,k),(l',k')} = -\int_{S^2} Q(\epsilon, x)[Y_l^k]Y_{l'}^{k'}\,\mathrm{d}\Omega = -\int_{\mathbb{S}^2} C_l Y_l^k Y_{l'}^{k'}\,\mathrm{d}\Omega = -C_l\delta_{(l,k),(l',k')}. \tag{13}$$

Spherical harmonics are eigenfunctions of isotropic operators, hence $\mathcal{C}$ is diagonal with entries C_l, which are the eigenvalues of the scattering operator Q corresponding only to the degree l of the spherical harmonic ψ_l^k. As for the stopping power, we derive additivity for the transport coefficients using the mass concentrations $\rho_i(x)$ from the additivity approximation [22]

$$C_l(\epsilon, x) = \sum_{i=1}^{n_e} \rho_i(x)\frac{C_{l,i}(\epsilon)}{A_{Z_i}} . \tag{14}$$

The specific coefficients $C_{l,i}(\epsilon)$ for the pure elements are derived from scattering cross-sections generated using the code from [25], which in [5] are tabulated for the different degrees of the spherical harmonics.

2.2.4. Advection matrices. By testing the advection operator $\Omega \cdot \nabla_x \Psi$ with the spherical harmonic $Y_{l'}^{k'}$, the advection matrices $A^{(x)}$, $A^{(y)}$ and $A^{(z)}$ are identified using the recursion relation given for spherical harmonics. Note that the advection matrices are very sparse and possess a specific pattern with respect to the classification of the spherical harmonics. Furthermore, all advection matrices are symmetric $A^{(n)T} = A^{(n)}$. The sparsity of the advection matrices is exploited in the numerical method and paves the way towards an efficient implementation of the P_N model.

2.2.5. P_N-boundary conditions. Initial conditions for the electron fluence Ψ in energy are specified at a given maximum energy ϵinit.

$$\Psi(\epsilon_{\text{init}}, x, \Omega) = \Psi_0(x, \Omega). \tag{15}$$

The translation of an initial condition in energy $\Psi_0(x, \Omega)$ into its moments $\psi_{l,0}^k(x)$ is straightforward. Using Eq. (10), the initial condition Ψ_0 is expanded into the respective moments. In the following experiments, we will typically use $\Psi_0(x, \Omega) = 0$, assuming that no beam electrons are present inside the sample at $\epsilon = \epsilon_{\text{init}}$.

For an implementation, the spatial variable x is confided to $\mathfrak{G} \subset \mathbb{R}^3$. Then a proper treatment of boundary conditions in space is necessary. At a point $x \in \partial\mathfrak{G}$ with n the outward boundary normal, we are allowed to prescribe the ingoing part of the electron fluence Ψ

$$\Psi(\epsilon, x, \Omega) = \Psi_{\text{in}}(\epsilon, x, \Omega) \quad \Omega \cdot n < 0 . \tag{16}$$

The boundary conditions for the moments ψ are given by Galerkin projections of Eq. (16) with spherical harmonics, which are odd functions with respect to the normal direction on S^2. See [18] for more details.

2.2.6 Beam parameters. The electron beam hits the sample at a surface (in our examples usually $\{(x_1, x_2, x_3) \in \mathfrak{G} | x_1 = 0\}$). We model the beam using an isotropic Gaussian distribution on the material surface, a Gaussian distribution in energy and a three-dimensional Von-Mises-Fisher distribution (a distribution defined on S^2 [26]) in direction. Using respective means μ_x and μ_ϵ, variances σ_x and σ_ϵ as well as the mean direction μ_Ω and concentration parameter κ of the Von-Mises-Fisher distribution, we define

$$\Psi_{\text{in}}(\epsilon, x, \Omega) = \mathcal{N}\left((x_2, x_3)^T | \mu_x, \text{diag}(\sigma_x)\right) \mathcal{N}(\epsilon|\mu_\epsilon, \sigma_\epsilon) \mathcal{F}(\Omega|\mu_\Omega, \kappa), \tag{17}$$

where $\mathcal{N}$ is the probability density function of the Gaussian distribution and $\mathcal{F}$ is the probability density function of the Von-Mises-Fisher distribution. Integration of $\mathcal{F}$ into the respective moments $\int_{n\cdot\Omega<0} \Psi_{in} Y_l^k(\Omega) d\Omega$ is performed numerically.

2.3. Numerical implementation

Computing k-ratios using the model presented above is based on STARMAP, a numerical method for the solution of the moment expansion for the electron fluence, based on a generic moment equation and detail on the solution of the P_N-equation. The main characteristic of STARMAP is the discretisation of the solution variable on staggered grids. Allowed by a specific structure of the equation, the staggered grid discretisation leads to an efficient second order approximation. It is originally published in [27] and already used in the context of electron transport in EPMA in [5, 18, 19].

3. Material representation

The particular choice of a model of mass concentrations ρ is problem dependent and should be conducted with great care and a particular material in mind.

In this section we provide models for mass concentrations which are utilised to conduct the reconstruction experiments presented in this paper. Thereby we distinguish between mass concentration models and parametrisations. The former is the relation of mass concentrations with other physical quantities and the latter is the mathematical parametrisation of a function in 3D. Combining a mass concentration model with a parametrisation yields the relation

$$\rho(x; p) : \mathbb{R}^3 \times \mathbb{R}^{n_p} \rightarrow \mathbb{R}^{n_e}, \tag{18}$$

that is, a parametrisation of the function $\rho(x)$, which returns mass concentrations of n_e elements given a position x inside the material and n_p parameters, which describe the material.

3.1. Parametrization models

In this section we propose some parametrisations for the discretisation of a multivariate 3D function. Although we visualise the concepts only in 2D, the extension to 3D or the reduction to 1D is straightforward. We denote the parametrised function with $P(x; p)$: $\mathbb{R}^3 \times \mathbb{R}^{n_p} \rightarrow \mathbb{R}^{n_o}$ and its parameters with $p \in \mathbb{R}^{n_p}$. We denote the number of output values of the parametrisation P with n_o. For the sake of comparability, we precisely formulate the number of required parameters n_p for each parametrisation.

3.1.1. Piecewise-constant parametrisation. Assume that the reconstruction domain is split into several cuboids(3D) / rectangles(2D), see Figure 2a, with aligned interfaces in all dimensions. We specify a rectangular box by

$$\mathcal{B}_{i,j,k} = \{(x, y, z) \in \mathbb{R}^3 | x_i < x \le x_{i+1}, y_j < y \le y_{j+1}, z_k < z \le z_{k+1}\} \tag{19}$$

where the interfaces are given by $\{x_i\}_{i=1,\dots,n_x}$, $\{y_j\}_{j=1,\dots,n_y}$ and $\{z_k\}_{k=1,\dots,n_z}$ with n_x, n_y and n_z the number of interfaces in each dimension. Based on the subsets, we define the 3D piecewise-constant parametrisation by

$$P(x; p) = \begin{cases} p_{i,j,k} & x \in \mathcal{B}_{i,j,k} \\ p_{\text{out}} & \text{else} \end{cases}. \tag{20}$$

In each of the boxes $\mathcal{B}_{i,j,k}$ the parametrisation P is constant with a value defined by the parameters $p_{i,j,k}$. If x happens to lie outside all the boxes, we assign p_{out}. The number of parameters is given by $n_p = ((n_x - 1)(n_y - 1)(n_z - 1) + 1)n_o$. Because $pout$ counts as an additional "box", we add 1. Note that the positions of the interfaces x_i, y_j and z_k are fixed and not part of the parameters.

3.1.2. Linear parametrisation. The linear parametrisation is also based in the splitting of the reconstruction domain into several boxes, see Figure 2b. But instead of specifying the value of P inside each box, we specify the value of P on the vertices of the boxes. We define the linear interpolation by

$$P(x,p) = \begin{cases} \mathcal{I}(x; p_{i,j,k}, p_{i+1,j,j}, p_{i,j+1,k}, \ldots, p_{i+1,j+1,k+1}) & x \in \mathcal{B}_{i,j,k} \\ \mathcal{I}(x; p_{\text{out}}) & else \end{cases} \tag{21}$$

where I denotes 3D linear interpolation of the values on the vertices.

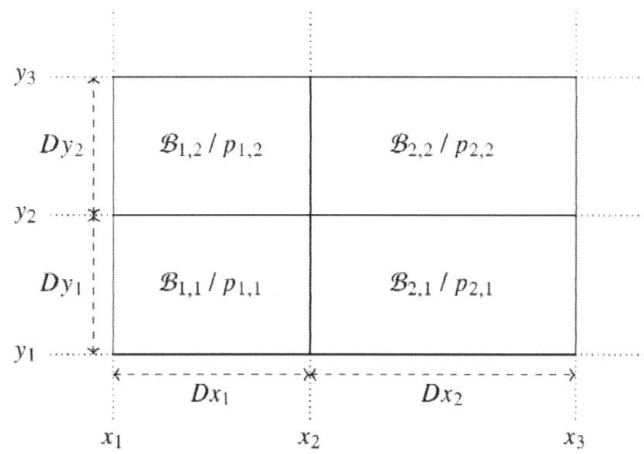

a) 2D sketch of the piecewise-constant parametrisation

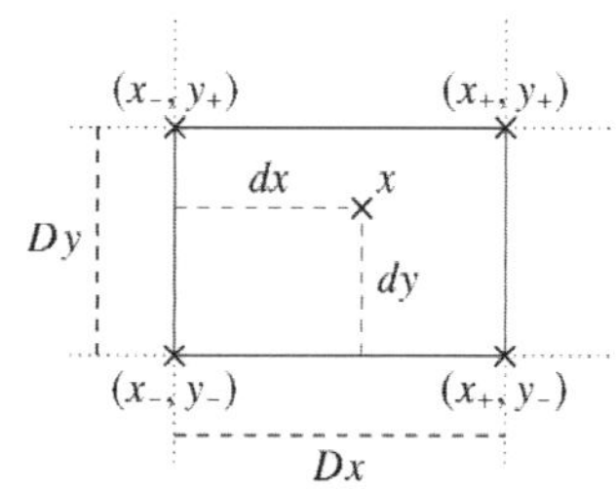

b) 2D sketch of the linear parametrisation

Figure 2. In a) interfaces of the boxes $\mathcal{B}_{i,j,}$ are defined by x_i and y_j. The boxes are not required to be equally sized, and their side lengths are defined by Dx_i and Dy_i. Inside each box $\mathcal{B}_{i,j,}$ the function value is constant $P(x \in \mathcal{B}_{i,j}) = p_{i,j}$. b) Shows a box with vertices (x_-, y_-), (x_+, y_-), (x_+, y_+) and (x_-, y_+). Distances of x from the interfaces are denoted with dx respectively dy and the side lengths of the box with Dx and Dy.

We visualise the linear interpolation in 2D in Figure 2b and derive the interpolation formula. Given a point (x,y) which is inside the Rectangle $\{(x,y)|x_- \leq x < x_+, y_- \leq y < y_+\}$, we derive interpolation weights η_x and η_y

$$\eta_x = \frac{dx}{Dx} = \frac{x - x_-}{x_+ - x_-} \quad \eta_y = \frac{dy}{Dy} = \frac{y - y_-}{y_+ - y_-}. \tag{22}$$

Based on the parameters defined on the vertices p_{--}, p_{+-}, p_{-+} and p_{++} the interpolation is given by

$$\mathcal{I}(x; p_{--}, p_{+-}, p_{-+}, p_{++}) = (1-\eta_x)(1-\eta_y)p_{--} + \eta_x(1-\eta_y)p_{+-} + (1-\eta_x)\eta_y p_{-+} + \eta_x\eta_y p_{++}. \tag{23}$$

A linear interpolation using n_x, n_y and n_z interfaces has $n_p = (n_x n_y n_z + 1)n_o$ parameters. We add 1 if we additionally specify a value for points outside all interfaces.

4. Reconstruction as an optimisation problem

The formulation of the reconstruction as an optimisation problem mainly consists of the following steps: The definition of the parameters p which are searched for, the definition of the forward model $k(p)$ and the definition of the objective

$$\mathrm{disc}(k(p), k^{\mathrm{exp}}) \tag{24}$$

as some measure of discrepancy between modelled $k(p)$ and measured k^{exp} k-ratios. The reconstruction result is then the set of parameters p^* for which the modelled k-ratios have the smallest discrepancy from the measured k-ratios

$$p^* = \arg\min_p \mathrm{disc}(k(p), k^{\mathrm{exp}}) \,. \tag{25}$$

Usually the discrepancy disc(·,) denotes a norm, therefore $\mathrm{disc}(\cdot,\cdot) \geq 0$. However, we cannot expect to find p^* such that $\mathrm{disc}(k(p^*), k^{\mathrm{exp}}) = 0$ in a real reconstruction problem, because of imperfect modelling and the presence of noise. Here we define the discrepancy $\mathrm{disc}(\cdot,\cdot)$ as the squared error of modelled and measured k-ratios

$$\mathrm{disc}(k(p), k^{\mathrm{exp}}) = \sum_{(Z,j)} (k_{(Z,j)}(p) - k^{\mathrm{exp}}_{(Z,j)})^2 \,. \tag{26}$$

4.1. Statistical interpretation

Defining the squared error as the objective function is very common, but may feel arbitrary. Inverse problem theory justifies the choice of the squared error function by deriving it from statistical assumptions [28]. The assumptions simultaneously allow a statistical interpretation of the reconstruction result p^*.

Inverse problem theory [28, 29] relies on the assumption, that the observation (the k-ratio measurements k^{exp}) relate with the model parameters p through

$$k^{\mathrm{exp}} = k(p) + \eta \,, \tag{27}$$

where η is a random variable. More generally, one defines the likelihood, the probability that based on the parameters p the model $k(p)$ produces the data k^{exp}. The probability density function (pdf) of the likelihood is

$$\pi(k^{\mathrm{exp}}|p) = \pi_\eta(k^{\mathrm{exp}} - k(p)), \tag{28}$$

where π_η is the pdf of the random variable η, which is usually referred to as noise, but might also include model uncertainties.

Additionally, the prior knowledge about the material is encoded in the prior pdf $\pi(p)$. Using Bayes Theorem, the resulting uncertainty about the material is the posterior pdf

$$\pi(p|k^{\mathrm{exp}}) = \frac{\pi(k^{\mathrm{exp}}|p)\pi(p)}{\int \pi(k^{\mathrm{exp}}|p)\pi(p)\,\mathrm{d}p} \,. \tag{29}$$

The posterior is the probability, that p represents the reality given the measured observation k^{exp}. The posterior $\pi(p|k^{\mathrm{exp}})$ combines likelihood $\pi(k^{\mathrm{exp}}|p)$ and prior $\pi(p)$ and defines the solution of the inverse problem. The statistical approach to inverse problems is philosophical and for problems of a certain size not (yet) applicable. However, it motivates concepts which are applied to solve real world reconstruction problems.

The maximum likelihood estimate p^* for EPMA is

$$p^* = \arg\max_p \pi(k^{\text{exp}}|p), \tag{30}$$

the maximum of the probability density function of the likelihood Eq. (28). Analogously we can write

$$p^* = \arg\min_p \underbrace{-\pi(k^{\text{exp}}|p)}_{=\text{disc}(k(p),k^{\text{exp}})}, \tag{31}$$

and define a discrepancy function. The maximum posterior estimate can be found similarly by defining

$$\text{disc}(k(p), k^{\text{exp}}) = -\pi(p|k^{\text{exp}}) \propto -\pi(k^{\text{exp}}|p)\pi(p), \tag{32}$$

where $\pi(p|k^{\text{exp}})$ is the pdf of the posterior, which is proportional to the product of the likelihood pdf and the prior pdf $\pi(p)$.

In the special case of a Gaussian likelihood and a Gaussian prior, the problem to find the maximum posterior can be written as

$$p^* = \arg\min_p ||k(p) - k^{\text{exp}}||^2_{\Sigma_L^{-1}} + ||p - \mu_P||^2_{\Sigma_P^{-1}}, \tag{33}$$

where Σ_L and Σ_P are the covariance matrices of likelihood and prior and μ_P the mean of the prior. With $||x||_A^2 = x^T Ax$ we denote weighted least squares. The special case of the maximum posterior approximation with Gaussian probabilities is also known as Tikhonov regularisation, a well-studied regularisation technique [1]. When neglecting the regularisation term $||p - \mu_P||^2_{\Sigma_P^{-1}}$ and choosing an isotropic covariance $\Sigma_L = \sigma_L^2 I$, Eq. (33) is equivalent to minimising the squared error Eq. (26).

A justification of the Gaussian assumptions or the definition of a proper likelihood and prior is beyond the scope of this work, because it requires a detailed uncertainty quantification of modelling errors and measurement noise. Furthermore, it is unclear if a general definition of the objective function for reconstruction in EPMA is possible, because the inverse problem is so multifaceted. In this paper, we employ the squared error as the discrepancy function.

5. Showcase: 1D reconstruction of interfaces

We present the reconstruction of a coated material consisting of iron and nickel with a sharp (discontinuous) and a diffusive (continuous) interface between the layers. By means of this example we will compare the piecewise-linear and piecewise-constant parametrisations described above. For a fair comparison, the number of parameters in each parametrisation is chosen to be 10. For the piecewise-constant parametrisation, the interval [0 nm, -300 nm] is divided in depth into 10 equal parts. Note that the location of the sharp interface aligns with one of the interfaces of the piecewise-constant parametrisation. For the linear interpolation, we discretise the interval in 9 equal parts. We strongly enforce the constraint for the volume fractions $\varphi_{Ni} = 1 - \varphi_{Fe}$, hence at every location a scalar $P(x; p)$ completely describes the material. To resolve the interface position and structure, we compare k-ratios from multiple different beam energies between 9 keV and 15 keV. Secondary fluorescence effects are neglected. Additional P_N-model settings and the considered X-rays are tabulated in Table 1. We use the squared error of k-ratios as the objective function and choose the BFGS (Broyden-Fletcher-Goldfarb-Shanno) [30] method implemented in OPTIM.JL [31] as optimisation method.

Table 1. Model settings used for the reconstruction of the 1D interfaces.

domain $\mathfrak{S}$	$(-500\text{nm}, 0\text{nm})$ $n_x = 100$
energies $[\epsilon_{\text{cut}}, \epsilon_{\text{min}}]$	$(7.0\text{keV}, 15.5\text{keV})$
approximation order P_N	$N = 9$
beam $\epsilon, \sigma_\epsilon$	$\{9, 10.5, 12, 13.5, 15\}\text{keV}, 0.1\text{keV}$
elements/x-rays	$(Ni, K - L_2)$, $(Ni, K - L_3)$, $(Fe, K - L_2)$ and $(Fe, K - L_3)$

In Figure 3, the reconstructed density of the material with a sharp interface is visualised using the different parametrisations. From initial configurations (piecewise-constant and -linear: 0.5 Fe and 0.5 Ni) the parametrisations iterate towards the reference density (black line) during the optimisation. We visualise the initial, first, fifth iteration and the 100th iteration. The inability of the linear parametrisation to represent discontinuities becomes apparent. However, the piecewise-constant parametrisation only approximates perfectly because the interface matches the interface of the reference.

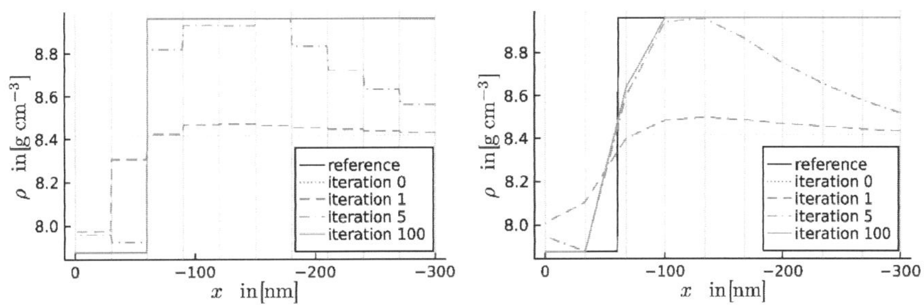

a) Piecewise-constant b) Piecewise-linear

Figure 3. The reconstructed density of a material with a sharp interface between an iron layer covering an nickel-substrate. We use the material parametrisations based on piecewise linear and piecewise-constant functions for the reconstruction. The geometry (interfaces) are visualised by grey vertical lines. The black line shows the reference density. All parametrisations converge. The piecewise-constant parametrisation has a clear advantage, because the interface aligns with an interface of the parametrisation. The linear parametrisation cannot approximate the discontinuous interface, hence it converges to the presented smoothed representation.

In Figure 4 the reconstructed density of the material with a diffusive interface is visualised. Except for the reference material, exactly the same settings as for the reconstruction of the sharp interface are used. Again, the parametrisations iterate towards the reference density (black line). For the diffusive interface, none of the parametrisations can reconstruct the interface perfectly, but the linear performs better than the piecewise-constant parametrisation.

In Figure 5 we visualise the normalised error, the value of the objective function $\|k(p) - k^{\text{exp}}\|^2$, the error in the mass concentrations $\|\rho(x) - \rho^{\text{true}}\|^2$ and the error of additional k-ratio measurements $\|k_+(p) - k_+^{\text{exp}}\|^2$ (with beam energies 11 keV and 14 keV) during the optimisation iterations. The kinks in the error of the objective function are an artefact of the OPTIM.JL implementation. Their default implementation is based on the combination of a line search method [32] and the BFGS algorithm [30].

IOP Conf. Series: Materials Science and Engineering **1324** (2025) 012005 doi:10.1088/1757-899X/1324/1/012005

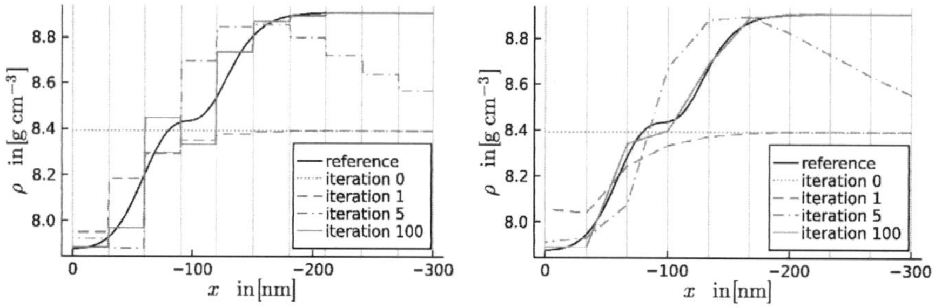

a) Piecewise-constant b) Piecewise-linear

Figure 4. The reconstructed density of a material with a diffusive interface between an iron layer covering an nickel-substrate. We again use piecewise-linear and -constant functions for the material parametrisations. The geometry (interfaces) are visualised by grey vertical lines. The black line shows the reference density. All parametrisations converge. None of the parametrisations can represent the interface perfectly. The bilinear performs better for this example, because of its flexibility to approximate continuous functions.

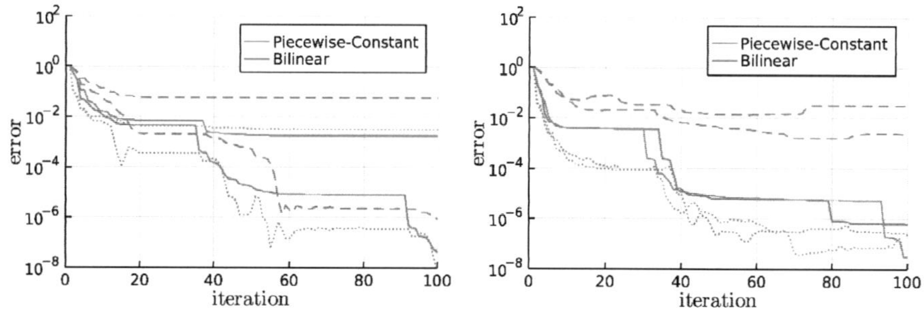

a) Sharp interface b) Diffusive interface

Figure 5. The value of the objective function and additional reconstruction quality measures during the iterations of the optimisation for all parametrisations. The objective function $\|k(p) - k^{\text{exp}}\|^2$ is visualised as a solid line. Also, the material error $\|\rho(p) - \rho^{\text{true}}\|^2$ (dashed line) and the error of k-ratios $\|k_+(p) - k_+^{\text{exp}}\|^2$ (dotted line) from additional beam energies (11 keV and 14 keV) is plotted.

Differences in the performance of the different parametrisations are clearly visible. The material error for the piecewise-constant parametrisation for the sharp interface is by far the smallest, due to the possible perfect reconstruction. Comparing the material error for piecewise-constant and -linear parametrisation for the sharp and the diffusive interface, they vary. A correlation between the material error and the error of additional k-ratios would be beneficial, but is hard to obtain. Ideally both would behave similarly, because then we could propose the error in additional k-ratios as a suitable reconstruction quality measure.

IOP Conf. Series: Materials Science and Engineering **1324** (2025) 012005 doi:10.1088/1757-899X/1324/1/012005

6. Showcase: 2D reconstruction of an ellipsoidal structure

We describe the use case of a reconstruction of an ellipsoidal copper inclusion in an iron-substrate. With this example we also show the extensibility of our method to an additional material parametrisation.

We are investigating a material consisting of Cu and Fe, where we simulate a measurement using a 12 keV electron beam with multiple beam positions. From a line-scan we observe the (non-noisy) *k*-ratio profile plotted in Figure 6. Only the *k*-ratios retrieved from five beam positions μ_y = {-50, -25, 0, 25, 50}nm which are marked by black crosses are used for the reconstruction. The profile shows an increase in the (Cu, *K*−*L2*) *k*-ratio for a beam position close to ≈ 25 nm. We assume that the material has elliptical Cu inclusions of different size, shape and interface structure.

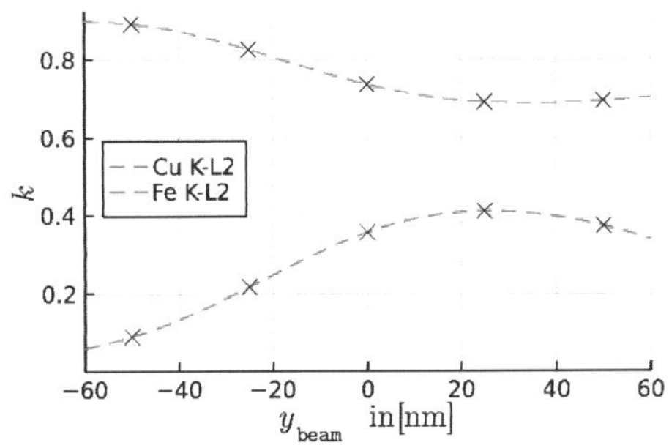

Figure 6. The *k*-ratio profile of the elliptical Cu inclusion in the Fe-substrate computed using a 12 keV electron beam. Dashed lines show the *k*-ratio profiles of (Cu, *K*−*L2*) (blue) and (Fe, *K*−*L2*) (orange) with a high beam resolution. Black marker show the *k*-ratios used for the reconstruction. While not being directly visible, the shape and height of the *k*-ratio curves encode information about the shape and location of the inclusion and the structure of the interface between inclusion and substrate.

We specify a material parametrisation which encodes the knowledge of an elliptical structure (with unknown position (μ_1, μ_2), unknown rotation r and unknown scaling factors for the principal axes (a, b)). For the unknown interface structure we use a non-linear sigmoid function. To complete the material parametrisation, we assume that we can model the mass concentrations $\rho_{Cu} = \varphi_{Cu}\rho_{Cu}^{pure}$ and $\rho_{Fe} = \varphi_{Fe}\rho_{Fe}^{pure}$ using volume fractions φ_{Cu} and φ_{Fe}, where $\varphi_{Cu} + \varphi_{Fe} = 1.0$, and densities of the pure materials $\rho_{Cu/Fe}^{pure}$. Hence, it is sufficient to parametrise the scalar volume fraction $\varphi_{Cu}(x)$ and deduce the volume fraction of Fe from $\varphi_{Fe}(x) = 1 - \varphi_{Cu}(x)$. As an initial guess, we assume that the inclusion is circular and that it is located around the maximum of the (Cu, *K*−*L2*) *k*-ratio profile.

The *k*-ratio shown in the line profile in Figure 6 are artificial (simulated) measurements. Hence, the reference material ρ^{true}, which is used to compute the artificial measurements k^{exp} is known. We visualise the total density ρ_{tot} of the reference material in Figure 7 and additionally show the ionisation distributions of (Cu, *K*−*L2*) and (Fe, *K*−*L2*) that define the size of the interaction volume. The ellipsoidal Cu inclusion is clearly smaller than the interaction volumes. Additional settings of the *k*-ratio model are shown in Table 2.

For the reconstruction we use the L-BFGS algorithm (limited-memory Broyden-Fletcher-Goldfarb-Shanno algorithm) [33] implemented in OPTIM.JL [31] using the default settings. The total density of the material ρ_{tot} during the iteration steps is visualised in Figure 8. From the initial guess the reconstruction converges to the reference material. Alongside the total density of the reconstructed material we visualise the *k*-ratio profiles of (Cu, *K*−*L2*) and (Fe, *K*−*L2*) in Figure 9. The shape of the curves quickly coincide with the measured *k*-ratios (black crosses). At iteration 30 the measured *k*-ratios are already well approximated, the total density, however, is not. For sufficient reconstruction of the ellipsoidal structure, the error in the *k*-ratios (the objective function) must be further reduced. In Figure 10 we show the (normalised) value of the objective function $||k(p) - k^{\text{exp}}||^2$ and the (normalised) material error $||\rho(p) - \rho^{\text{true}}||^2$ during the iterations of the optimisation. For the first

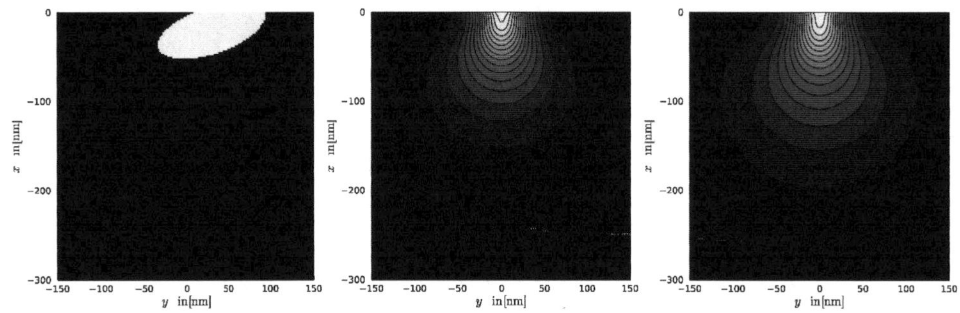

a) Reference density ρ_{tot}^{true} b) Ionisation $\phi_{(Cu,K-L_2)}$ c) Ionisation $\phi_{(Fe,K-L_2)}$

Figure 7. The total density and two ionisation distribution fields of the reference material. The ionisation distribution curves determine the size of the interaction volume. The size of the ellipsoidal Cu inclusion is smaller than the interaction volume.

Table 2. Model settings used for the 2D reconstruction of the ellipsoidal Cu inclusion.

domain $\mathfrak{S}$	$(-300\text{nm}, 0\text{nm}) \times (-150\text{nm}, 150\text{nm})$ $(n_x, n_y) = (80, 80)$
energies $[\epsilon_{\text{cut}}, \epsilon_{\text{min}}]$	(7.0keV, 12.5keV)
approximation order P_N	$N = 9$
beam $\mu_\epsilon, \sigma_\epsilon$	12keV, 0.2keV
beam μ_x, σ_x	$(0, \{-50, -25, 0, 25, 50\})$nm, 20nm
beam μ_Ω, κ	$[-1, 0]$, 10
elements/x-rays	$(Cu, K-L_2)$, $(Cu, K-L_3)$, $(Fe, K-L_2)$ and $(Fe, K-L_3)$
standard materials	$(Cu, K-L_*) : (1.0Cu, 0.0Fe)$, $(Fe, K-L_*) : (0.0Cu, 1.0Fe)$

50 iterations, the value of the objective function decreases, whereas the material error remains close to the initial error. Only after the first 50 iterations, the material error also decreases. In Figure 8 this effect is also observed, where only after iteration 60 the shape of the reference ellipse starts to form.

6.1. Comparison: Representation capabilities of the material parametrisations

We compare the capability of the piecewise-constant and the bilinear parametrisation to represent the ellipsoidal material structure. Therefore, the material error $||\rho(x;p) - \rho^{\text{true}}(x)||^2$ is optimised directly. Piecewise-constant and -linear parametrisation only discretise a section of the visualised domain in Figure 7a. For a fair comparison, the number of parameters n_p of each parametrisation is 144. For the piecewise-constant representation we use (−100 nm, 0 nm) × (−120 nm, 120 nm) with (n_x = 13, n_y = 3, n_z = 2), while the piecewise linear case is based on (−100 nm, 0 nm) × (−120 nm, 120 nm) with (n_x = 12, n_y = 12, n_z = 1). Figure 11 shows the results of an optimisation using the L-BFGS method implemented in OPTIM.JL. Clearly, artifacts of the parametrisations are visible. The piecewise-constant parametrisation cannot approximate the sharp interface of the ellipsoidal inclusion, and averages in boxes which intersect the interface. Also, the linear parametrisation cannot approximate the sharp interface, but represents the interface by a gradient. Both parametrisations would perform better, if the number of parameters would be higher.

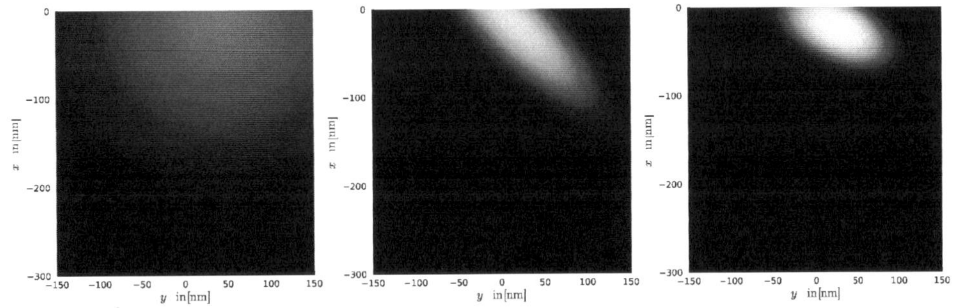

a) Initial b) Iteration 30 c) Iteration 60

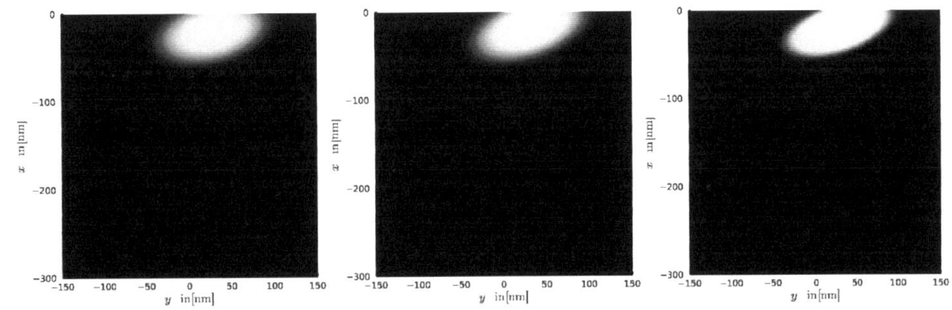

d) Iteration 90 e) Iteration 120 f) Iteration 250

Figure 8. The reconstructed density ρ_{tot} of the Fe material with a Cu inclusion during the iterations of the L-BFGS optimisation.

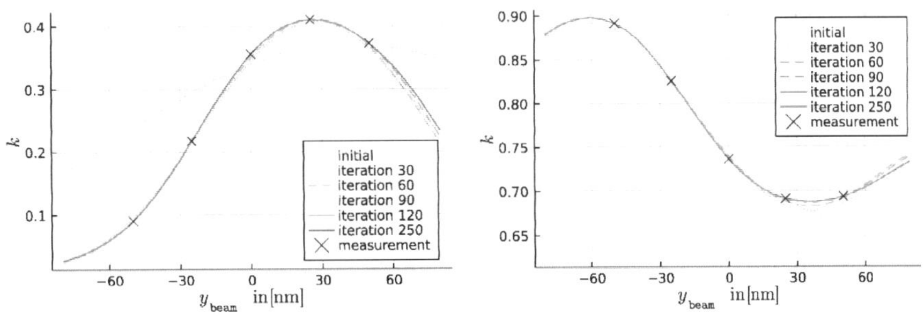

Figure 9. The k-ratio profile of (Cu, $K{-}L2$) and (Fe, $K{-}L2$) during the iterations of the reconstruction. Black crosses mark the measured k-ratios, which are considered in the objective function $\|k(p) - k^{\text{exp}}\|^2$ (c.f. Figure 6). The visualised iterations are the same as in Figure 8. Note that the k-ratios quickly converge to the measured values, although the shape of the ellipse does not yet coincide with the shape of the ellipse in the true material (Figure 7).

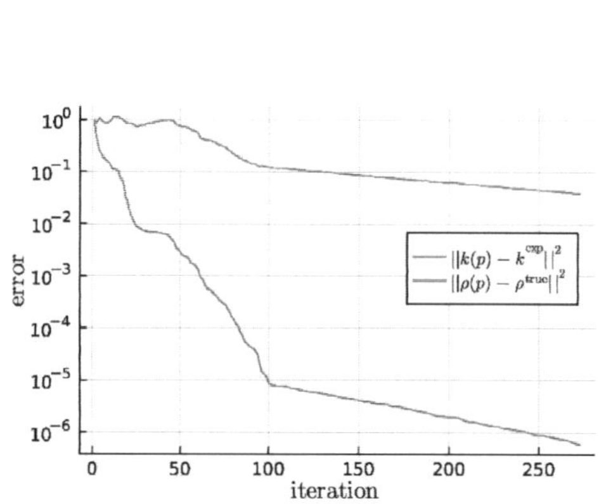

Figure 10. Normalised errors of the objective function $||k(p) - k^{\exp}||^2$ and the material error $||\rho(p) - \rho^{\text{true}}||^2$ during the iterations of the L-BFGS optimisation. Although the objective function significantly decreases for the first 50 iterations, the material error remains close to the initial error.

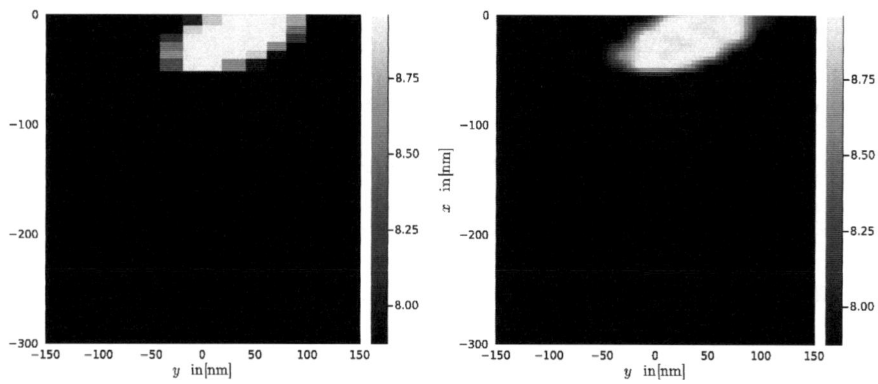

Figure 11. Resulting total density $\rho_{\text{tot}}(x)$ of the direct minimisation of the material error $||\rho(x;p) - \rho^{\text{true}}(x)||^2$. They allow a comparison the representation capabilities of the 2D parametrisations. The reference $\rho^{\text{true}}(x)$ is the ellipsoidal inclusion from Figure 7a. Clearly artifacts of the parametrisations are visible.

7. Conclusion

Deterministic transport equations offer an alternative to Monte Carlo methods for modelling electron transport in inhomogeneous materials. Furthermore, the deterministic P_N-model is well suited for reconstruction due to its efficient implementation and differentiability. Examples presented in this paper indicate, that sub-scale inversion, i.e., the reconstruction of material structures which are smaller than the interaction volume, is possible with sufficient data or/and sufficient material assumptions. But there will always be a relation between the accuracy of the measurement and the model to the accuracy of the reconstruction. Understanding the uncertainty that measured k-ratios as well as approximations in the model propagate into the reconstruction result is key to a robust reconstruction.

8. Acknowledgments

The co-authors Gaurav Achuda and Tamme Claus are funded by the German Research Foundation (DFG) through the individual research grant with project number DFG-466010736.

References

[1] Benning M and Burger M 2018 *Acta Numerica* **27** 1-111
[2] Larsen E W, Miften M M, Fraass B A and Bruinvis I A 1997 *Med. Phys.* **24** 111-125
[3] Duclous R, Dubroca B and Frank M 2010 *Phys. Med. Biol.* **55** 3843-3857
[4] Mevenkamp N 2013 Master thesis. [Aachen, Germany: RWTH Aachen]
[5] Bünger J 2021 *Three-dimensional modelling of X-ray emission in electron probe microanalysis based on deterministic transport equations.* PhD thesis. [Aachen, Germany: RWTH Aachen University] Publ. No. RWTH-2021-05180
[6] Naumann U 2011 *The art of differentiating computer programs software, environments and tools.* [Philadelphia, PA: Society for Industrial and Applied Mathematics]
[7] Griewank A 2003 *Acta Numerica* **12** 321-398
[8] Plessix R E 2006 *Geophys. J. Int.* **167** 495-503
[9] Pinard P T 2016 *Electron probe microanalysis of carbon containing steels at a high spatial resolution.* PhD.thesis. [Aachen, Germany: RWTH Aachen University]
[10] Ritchie N 2021 *FFAST.JL.* gitHub Repository URL https://github.com/usnistgov/FFAST.jl
[11] Ritchie N 2021 *NEXLCORE.JL.* gitHub Repository URL https://github.com/usnistgov/NeXLCore.jl
[12] Chantler C, Olsen K, Dragoset R, Chang J, Kishore A, Kotochigova S and Zucker D 2009 *X-ray form factor, attenuation, and scattering tables.* URL http://physics.nist.gov/ffast
[13] Ritchie N 2020 *BOTESALVATICX.JL.* gitHub Repository URL https://github.com/usnistgov/BoteSalvatICX.jl
[14] Bote D, Salvat F, Jablonski A and Powell C J 2009 *Atom. Data Nucl. Data Tables* **95** 871-909
[15] Bote D and Salvat F 2008 *Phys. Rev. A* **77** 042701
[16] Perkins S T, Cullen D E and Seltzer S M 1991 *Tables and graphs of electron-interaction cross-sections from 10 eV to 100 GeV derived from the LLNL Evaluated Electron Data Library (EEDL), Z = 1–100.* [Livermore, CA: Tech. Rep. Lawrence Livermore National Laboratory]
[17] Perkins S T, Cullen D E, Chen M H, Rathkopf J, Scofield J and Hubbell J H 1991 *Tables and graphs of atomic subshell and relaxation data derived from the LLNL Evaluated Atomic Data Library (EADL), Z = 1–100.* [Livermore, CA: Tech. Rep. Lawrence Livermore National Laboratory]
[18] Bünger J, Sarna N and Torrilhon M 2022 *J. Comput. Math.* **40** 979-1005
[19] Bünger J, Richter S and Torrilhon M 2022 *Microsc. Microanal.* **28** 454-468
[20] Cercignani C 1988 *The Boltzmann equation and its applications.* in: Applied Mathematical Sciences. [New York, NY: Springer-Verlag]
[21] Müller C 1966 *Spherical harmonics.* In: Lecture notes in mathematics. [Berlin Heidelberg, Germany: Springer-Verlag]
[22] Salvat F, Berger M J, Jablonski A, Bronic I K, Mitroy J, Powell C J and Sanche L 2007 *Elastic scattering of electrons and positrons.* ICRU Report 77 [Bethesda, MD: International Commission on Radiation Units & Measurements]
[23] Reimer L 1998 *Scanning electron microscopy: Physics of image formation and microanalysis. 2nd edition.* Springer Series in Optical Sciences. [Berlin Heidelberg, Germany: Springer-Verlag]
[24] Salvat F 2019 *PENELOPE 2018: A code system for Monte Carlo simulation of electron and photon transport.* [Barcelona, Spain: Nuclear Energy Agency]
[25] Salvat F, Jablonski A and Powell C J 2005 *Comput. Phys. Comm.* **165** 157-190
[26] Mardia K and Jupp P 2009 *Directional statistics.* Wiley Series in Probability and Statistics. [Chichester, England: Wiley]
[27] Seibold B and Frank M 2014 *ACM Trans. Math. Softw.* **41**
[28] Tarantola A 2005 *Inverse problem theory and methods for model parameter estimation.* Other Titles in Applied Mathematics. [Philadelphia, PA: Society for Industrial and Applied Mathematics]

[29] Stuart A M 2010 *Acta Numerica* **19** 451-559
[30] Nocedal J and Wright S 2006 *Numerical optimization. 2nd edition.* Springer Series in Operations Research and Financial Engineering. [New York, NY: Springer-Verlag]
[31] Mogensen P K and Riseth A N 2018 *J. Open Source Softw.* **3** 615
[32] Hager W W and Zhang H 2005 *SIAM J. Optimization* **16** 170-192
[33] Liu D C and Nocedal J 1989 *Mathem. Programming* **45** 503-528

Investigation of detector efficiency with EDS in SEM with the aim to improve standardless quantitative results

F Eggert, J Rafaelsen and P P Camus

AMETEK Materials Analysis Division, EDAX LLC
5794 West Los Positas Blvd., 94588 Pleasanton, California, U.S.A.

E-mail: frank.eggert.edx@outlook.de

Abstract. We have found deviations in in net-count measurement comparisons with EDS detectors equipped with different X-ray entrance windows that cannot be explained by the theoretical assumed and synchrotron measurements verified window transmissions. This work presents the results with systematic simultaneous measurements, which were performed with two EDS SDD systems on one SEM to verify this effect. The two detectors were equipped with a polymer window (Moxtek AP3) and the next generation Si_3N_4-window (Amptek C2), respectively. A possible explanation of measured detector-efficiency deviations will be given with an additional "divergence effect" due to the Moxtek polymer window support grid blades. It is to consider an additional X-ray transmission at window support grid lamella edges for divergent X-rays at energies already below 10 keV, while the support grid becomes significantly transparent in the parallel beam only with energies >10 keV. This effect is to consider in difference to the parallel beam transmission view because the X-rays in the SEM are from a point source, a fact which was ignored in the past. The "divergence effect" increases with detector area and with closer detector distance to the specimen in SEM. Small deviations of grid lamella alignment to the detector axis can produce similar lamella-edge transmission effects, which can differ between individual detectors. Taking the additional polymer window grid-lamella transmission into account explains the already observed and reported systematic difference in the calculated X-ray counts emitted from a specimen (measured net counts corrected with detector-efficiency). And with this, it explains the deviations of quantitative results between the two different detector window systems. Standardless quantitative results need to be independent from the detector (window) with which the measurements were performed. This was the motivation for the presented work and is a precondition to investigate and improve the standardless quantitative models further, to achieve improved accuracy in standardless evaluation of the measured X-ray data with energy-dispersive X-ray spectrometry (EDS) systems, used for the electron probe microanalysis (EPMA) method.

1. Introduction

The key measurement values and base for standardless quantitative analysis are the measured X-rays that were emitted from the specimen. A high degree of knowledge and control is required about the complete detector efficiency to determine these base values from the X-ray net-counts of the measured spectra. It is also a basis for assessment and improvement of the quantification models and parameters. For this, the measurement results need to be independent from the detector used and from all detector influences on measured results, to make a clear assessment of the X-ray production and absorption processes in analysed specimens. A significant part of the complete detector efficiency influence is the transmission of the atmospheric window. There are now two major types of windows used for scanning electron microscopy – energy-dispersive X-ray spectrometry (SEM-EDS) systems, the polymer window [1, 2] and the more recent Si_3N_4 window [3-5]. It is expected that windowless detectors will also be used in SEM applications in the future. This work deals with aspects that need to be considered and with the characterisation of the used detector windows for X-ray transmission in a typical SEM application. The detector efficiency effects are to clearly separate from the models and parameters, which are used to calculate the generation and absorption of X-rays inside the specimens until X-ray emission. It is a prework to making the analytical results consistent, independent of the used detector window, a basis to assess the X-ray production yields which are a basis to use the physical-based "fundamental parameter models" (like ZAF) also standardless with EDS-EPMA.

2. Problem

Because of the described importance, the energy-dependent transmission of X-rays through the used windows was typically measured at synchrotron facilities with a well collimated parallel X-ray beam [2, 4, 6]. But total detector efficiency is strongly correlated to the complete detector and window design [7], might also be different with non-parallel X-ray measurement, like it is always in SEM applications where the X-rays come from a point source.

Both types of X-ray windows are supported by a grid for mechanical stability. But the grids have different designs (Fig. 1). While the polymer window lamella grid has a 380 µm thickness of each rib (Fig. 2) [2], the "honeycomb" grid of silicon nitride window only has about 15 µm thickness (Fig. 3). The energy-dependent X-ray transmission curves are not only different at low energies, but also different in the important measurement region of EDS-EPMA between 3 - 10 keV, based on the very different designs and thickness parameters of the Si support grids [Fig. 4]. The polymer window grid with 380 µm thickness of the ribs is significantly transparent with energies >10 keV, while the new silicon nitride window grid with about 15 µm thickness becomes already transparent for lower energy X-rays.

Figure 1. SEM images showing the support grids for Moxtek AP3 polymer window (left) and silicon nitride window C2 by Amptek (right) [8].

	AP3
Thickness (µm)	380
Rib width (µm)	59
Opening width (µm)	190
Open Area %	76%
Acceptance Angle	53°

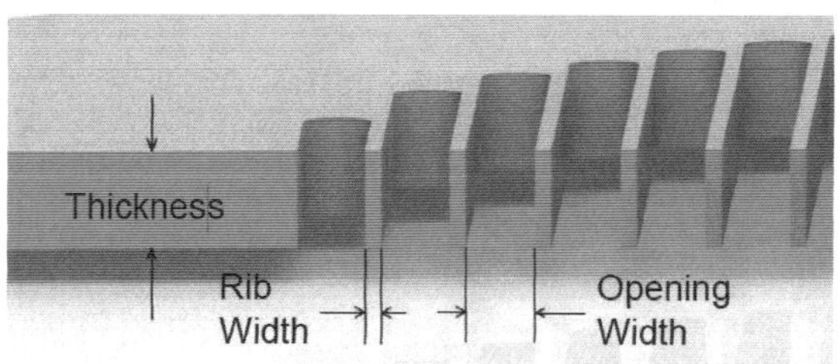

Figure 2. Design of Moxtek polymer window grid ribs (lamella) [2] with 380 µm rib thickness.

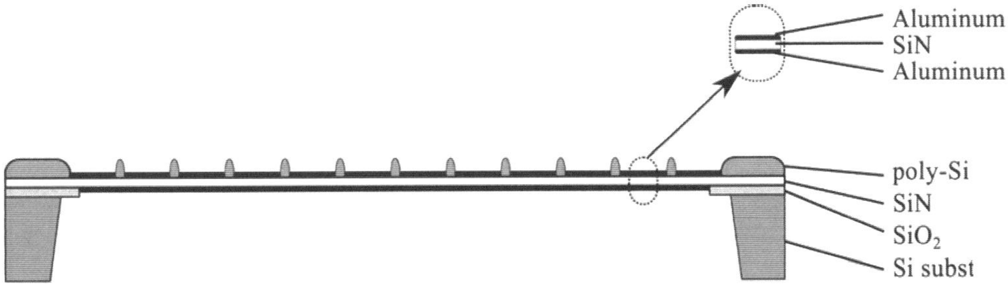

Figure 3. Design of Amptek C2 silicon nitride window grid [3] with about 15 μm rib thickness. The silicon nitride window is with a thickness of 40 nm with Al layers of 30 nm total. The open area is 80 %.

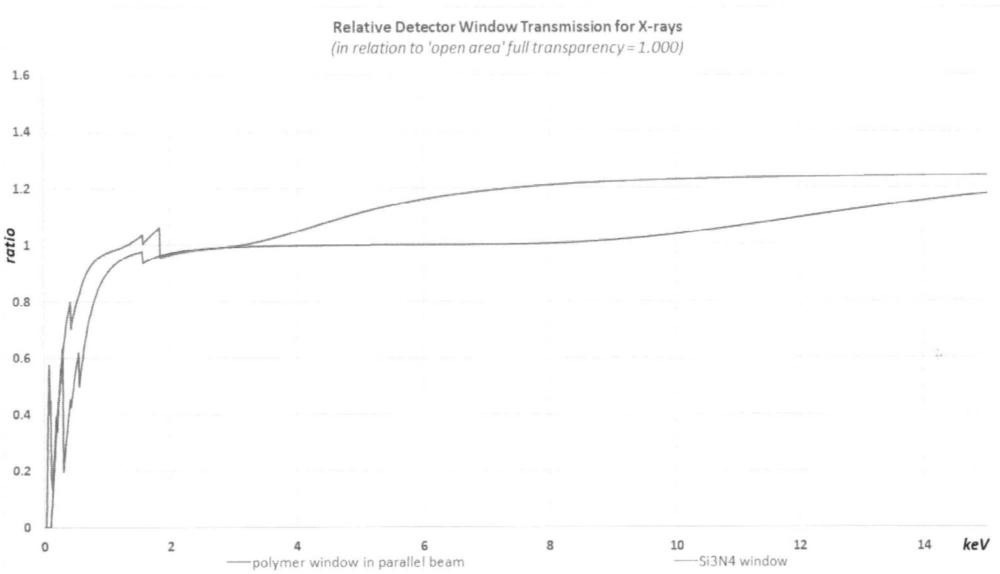

Figure 4. Comparison of the two window type transmissions calculated based on the known designs and parallel beam X-rays assumption. It was verified in synchrotron X-ray beam measurements [1, 4].

We have investigated the new type of Si_3N_4 window in SEM applications and have already found and discussed some measurement result inconsistencies with the expectation [9], if the commonly assumed transmission curves (see Fig. 4) are applied for X-ray spectra evaluations which were measured with EDS in SEM. The net count ratio differences with different X-ray lines, measured with the two different window types equipped detectors, were not possible to explain by the commonly used and assumed to be well known both windows transmissions [9]. It was already pointed out the reasons could be the X-ray point source in SEMs have different effects to the window support-grids transmission properties compared to parallel X-rays. While the silicon nitride thin "honeycomb" grid was already tested to have no or comparably very small transmission effects depending on X-ray incident angle [4], the polymer window lamella-grid must have effects, if the X-rays are not parallel to the grid lamellas. This is the case in SEM applications where the X-rays are divergent as more of the detector is close to the specimen surface and as the detector size is higher. The measured differences were explained as a 'divergence effect' of polymer detector windows if EDS detectors are used in the SEM.

3. Method and experimental

In this study, we show the first systematic efficiency comparison measurement between a polymer window and a silicon nitride window detector in the SEM. The idea was to determine the detected efficiency differences compared to theory by direct comparison of the same specimen measurements with a chain of pure elements samples. If the measured net-counts are corrected by the "known" assumptions with detector efficiency, based on the different "known" windows transmission curves, then there should be the same measured absolute concentrations based on the measured specimen X-ray emissions. Both detector-types (window-types) were operated in a parallel measurement setup. They have individual solid angle values, but the same beam-current was used. Both detectors' measurements were evaluated with same quantitative model, the same physics is valid, same sample was used and same excitation. The difference is the 'absolute adjustment' with both different detectors, which is a scalar value which represents the solid angle differences. And the second difference between both detectors is the energy-dependent detector efficiency, which is under investigation in this paper. The differences with absolute geometry for both detectors were considered by a factor for each detector, which was adjusted on the X-ray lines measured at 1 to about 2 keV X-ray energy. Both windows have similar transmission characteristics at this energy range and the support-grid transparency influences are expected to be minimal at this energy. If adjusting both measurements at other energy regions, it would only shift the Figure 5 results vertical. But the possibility to interpret the measurements which are shown in Figure 5, by physical reasons finally gives a strong additional support that the applied adjustment is reasonable.

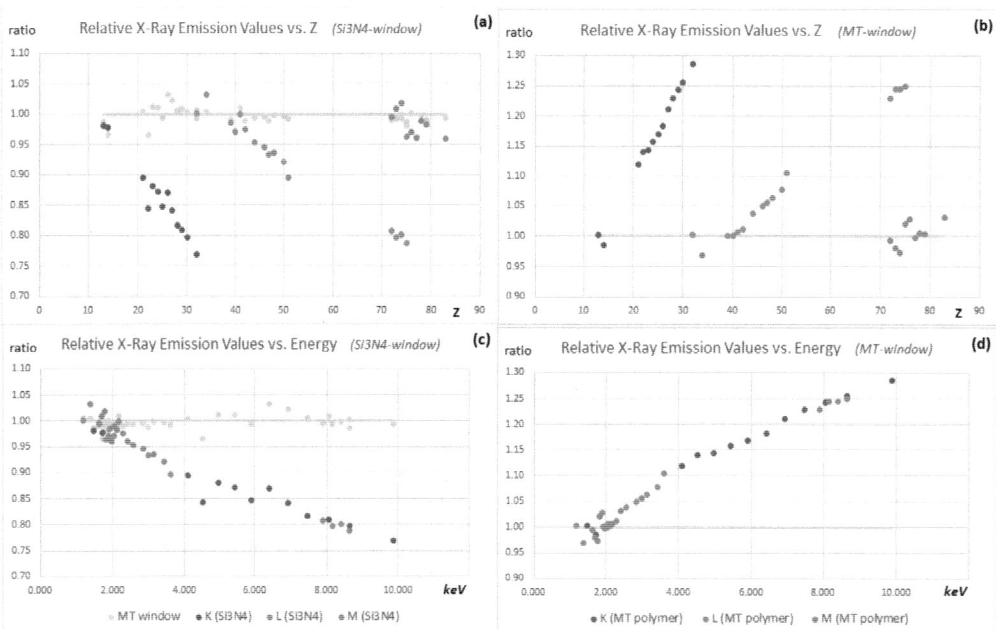

Figure 5. Measured standardless quantitative result deviations relative to "known" efficiency (pure element specimen; HV = 20 kV; absolute solid-angle adjustment around 1…2 keV energy region) versus atomic numbers of the elements and versus line energies. Left hand curves result-deviation plots from ideal 100 % concentration a) and c) are based on assumption, that the efficiency is already properly described with the traditional Moxtek (MT) polymer window. Right curves b) and d) with deviations of the polymer window assuming that the new Amptek C2 silicon nitride (Si_3N_4) window detector efficiency is true also in SEM measurements.

Figure 5 shows measured standardless quantitative result deviations which are equivalent to the deviations of the calculated X-rays emitted from the specimen (emitted net counts) with pure element specimen. The X-ray emission from the specimen was determined by correction of the net-counts (extracted from spectra evaluations) with the different efficiencies of the detector. The results at first view are not possible to explain with the two detector-window types. The deviations versus atomic number show a sawtooth profile in both views, left curves based on assumption that the old views with the polymer-window transmissions are proper (like the standardless model was used in past, before silicon nitride window introduction). With the right curves it is assumed that the new silicon nitride window transmission curves might be a new basis (and the X-ray emission yields are corrected then respectively, a revision of the old used database need). An improved view is the lower plot if the deviations are displayed versus X-ray energies. There is always a slope or decay over energy, the only possible explanation is with detector efficiency effects (because the emission for the parallel measurements was always the same from specimen surface with non-tilted specimen and same elevation angle). The left curve (Figure 5c) would require an explanation, that the new silicon nitride window detector needs to somehow have more absorption in the energy region > 2 keV than assumed, and more absorption than it was verified with synchrotron X-ray measurements. Alternatively, if the ratio of both measurements would be adjusted to 1.000 for 10 keV, a higher than assumed relative transmission in the low energy region could explain the measured curve. But this would only be possible to explain without physical assumptions with having the window even amplifying the X-rays with energies < 1 keV. The right curve (Figure 5d) is based on the same measurements, but assuming the new window transmission characteristic is the truth. It is with the result that the traditional polymer window must have higher transmission with > 2 keV than assumed until now. There is a possible explanation with the divergent X-rays which penetrate the polymer window lamella-grid partially at grid edges. This is not possible in the parallel X-ray beam case where the X-rays penetrate the lamella-grid in a significant manner only with energies higher than about 10 keV (see Figure 4).

We performed an extended systematic investigation with the dual-detector configuration in a SEM. Both detectors were mounted with identical elevation angels and intersection distances, making simultaneous data collection from the same area and with the same specimen X-ray path geometry possible. Everything was controlled under the "absolute" excitation view. Different element X-ray lines were compared with the absolute measured net counts between both detector window types, from pure element and additionally with some binary specimens. The EDAX Octane Plus (polymer window, Moxtek AP3) was compared with the EDAX Octane Elite Super (Si_3N_4 window, Amptek C2). The measurements were done for various X-ray lines with 5, 10, 15, 20, 25 and 30 kV SEM HV, with pure element and binary specimens. This provided us with an extended chain of experimental data from different X-ray line energies with the ability to assess the real energy dependent efficiency differences in an SEM application. All evaluations were based on net-counts measurements only. A commercially spectra processing and evaluation was used to extract the net-counts from the measured spectrum [10]. There was no quantitative matrix correction model used in these evaluations. The raw measured net counts were scaled by the detector efficiency based on the theoretical "known" window efficiency calculations (confirmed by synchrotron measurements), see Figure 4. But the polymer window theoretical transmission was used with zero grid transmission assumption (the transmission value stays 1.000 then also for energies > 8 keV). Figure 6 shows the measured relative deviations of corrected net counts from the polymer window divided by the new Si_3N_4 window (only the 15 and 30 kV measurements).

4. Results

The measured curve is interpreted as being due to X-ray transmission through the lamellas of the polymer window, caused by the divergent X-ray beam ("divergence-effect") [8], and additionally due to transmission at lamella edges, if the lamella alignment is not perfect. Due to the significantly thinner support grid and the design, the divergent beam effect is negligible for the Si_3N_4 window [4]. Consequently, it can be assumed by our investigation to be the "true" and known transmission for detector efficiency, also in the SEM application. The measured jump at the Si-K edge visible in Figure 6 clearly shows that the discrepancies are related to absorption in a Si material. The full line in Figure 6

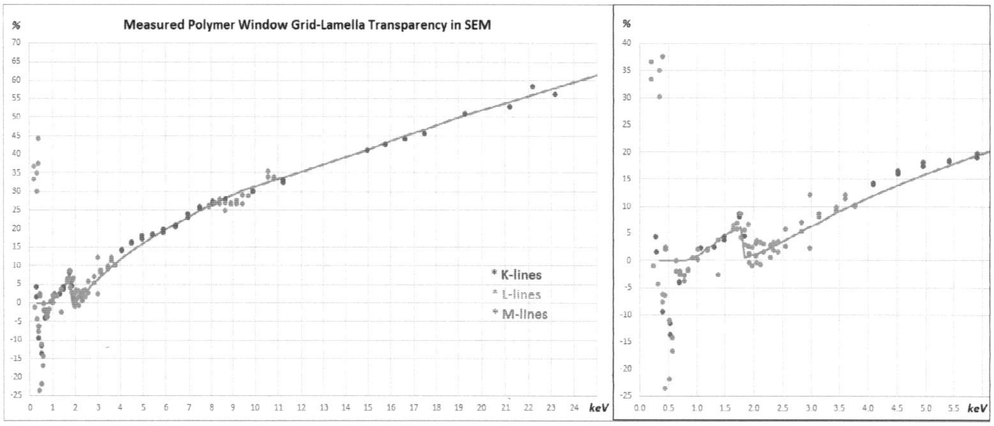

Figure 6. Measured net-counts of polymer window divided by measured net-counts of Si_3N_4-window (the 15 and 30 kV measurements only), both measurement results were corrected by "known" efficiency for parallel X-ray beam (the polymer window without considering the parallel beam grid transparency). The measured curve is, therefore, the complete grid transmission of the polymer window in the SEM with divergent beam and likely additionally for the parallel beam based extra-transmission due to an alignment deviation of the polymer window lamella compared to detector axis.

is a modelled curve that takes absorption and transmission through Si material into account based on five different X-ray paths. This approach models by an empirical way also the additional partial transmission through the edges of the MT-window grid-blades, due to the divergent X-ray beam and due to an additionally possible misalignment of the grid blades in relation to detector axis.

The "polymer window in SEM" line curve in Figure 7 represents the measured window transparency of the used detector with polymer window, considering the described extra transmission at lamella edges. The "polymer window parallel beam" dashed line is the transparency in the parallel beam with ideal grid blade alignment.

Figure 8 shows the repeated quantitative evaluations with similar 20 kV measurements like Figure 5, but all results now consider the new measured Moxtek AP3 polymer window transmission curve for SEM applications of EDS, considering also the "divergence effect". The new results do not show significant systematic deviations between measurement chains performed with the two detector-window design types, as it was before (see Figure 5). There are still accuracy deviations from expected values, especially below 2 keV, but for both detector types with similar deviations. This is a question remaining for further improvement of the applied quantitative models and parameters for electron and X-ray interactions in specimen, which was not yet the topic of this work.

5. Discussion

The goal was reached to get consistent and detector-independent quantitative results, the basis to improve now the models and databases for X-ray production and absorption in specimen (e.g., with ZAF correction). For this, a revision of the polymer window transmission properties in the SEM application of EDS was required. The detector effects on measurement results by window absorption and by limited detector-thickness based X-ray transmission were separated that way from the X-ray production and absorption effects in the specimen (the interaction processes by injected electrons and emitted X-rays).

As a result, the standardless quantitative results are now detector independent. The revision of window transmission view with the traditional Moxtek AP3 polymer window requires then vice versa the need to readjust the used atomic parameters in quantitative correction models for X-ray production in the specimen. All was done to use one model and basic database for the traditional window and also for the more recent silicon nitride window, and hopefully it will be still valid with windowless detectors in future.

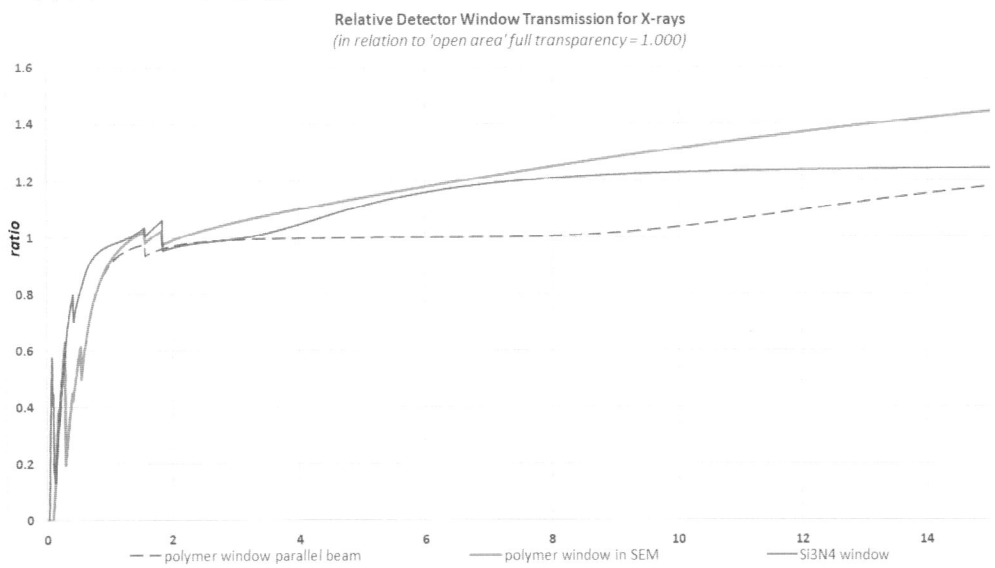

Figure 7. Comparison of two types of window transmission curves measured in SEM. The green "polymer window in SEM" (full line) is the new measured transmission of Moxtek AP3 polymer window with our parallel dual detector measurement, the "polymer window parallel beam" dashed line is the theoretical and verified transmission of this window type for parallel beam. The deviations of both lines are presenting the "divergence effect" of polymer window lamella-grid for point source X-ray measurements in SEM with an optional additional part of imperfect alignment of the grid-lamellas to the detector-axis and additionally the uncertainty with alignment of this axis to the point-source of X-rays on specimen (see measurement results in Figure 6).

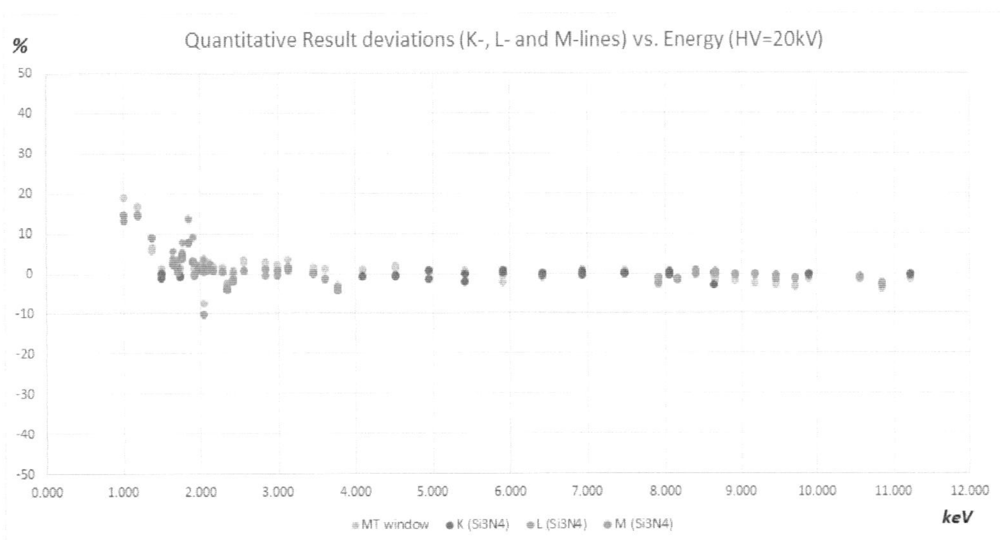

Figure 8. Improved quantitative results (compare to Figure 5d) after applying the divergence-effect modified transmission curve for Moxtek AP3 window. The grey dots are all results achieved with the Moxtek AP3-window detector measurement; the coloured dots are results measured with new Amptek C2 silicon nitride window-based EDS-detector.

IOP Conf. Series: Materials Science and Engineering **1324** (2025) 012006 doi:10.1088/1757-899X/1324/1/012006

The authors tried to calculate the "divergence effect" by numerically integration over geometry, based on the used detector size parameters and specimen-window distances. The results indicate the curve shape of Figure 6 in principle, but the measured effect is about 2 or 3 times higher than the theoretical calculations would predict. But the numerical calculations were with always ideal assumptions with the lamella shapes. Additionally, the "divergence effect" should vary with detector distance, should disappear as more of the detector is retracted. Several different measurements with moving the detector (with polymer window lamella grid) have provided a lower (or even no) effect than theoretical calculations predict. The reasons for both can be that real-world detectors with polymer window grid-lamellas also have some misalignments of the blades in relation to detector axis, which is in addition to the "divergence effect". The effect with variation of "divergence effect" via different detector distances is only possible to measure a clear way if the window support-lamella are ideally aligned to the detector axis, otherwise the effect is covered or much reduced. This lets us assume that there are also additional uncertainties in quantitative results due to the polymer window grid-lamella, it may have different effects with individual detectors. The misalignment between detector axis and real X-ray emission point on the specimen can also result in additional uncertainties and detector efficiency variations. This is then possible even between different measurements with the same detector, it would influence even standards-based analysis. The new silicon nitride thin "honeycomb" grid design does not have these additional geometry effect uncertainties in general, like those caused by a grid-lamella design.

In Figure 6 shown measured data of 15 and 30 kV excitation and the other HV not displayed measurement results, all clearly show that the efficiency determination gets unreliable for L- and M-lines below X-ray energies of about 600 eV. The reasons are the complex structure of L- and M-line series, which are with their energies often very close to the typical absorption edge jumps of low-Z window elements (C, N, O), which dominate the detector-efficiency. And even the individual lines with an L- or M-line series of a single element often have huge differences in absorption, if the window's element absorption-edges are between the line energies. While the K-lines below 600 eV are usually always at the absorption minimum, the L- and M-lines are very close to the window elements X-ray absorption jump energies, where the MACs (mass absorption coefficients) are very uncertain in this energy region. It is a further indication against the low-kV analytical method, which frequently requires the use of these low-energy L- and M-lines of the elements. It is expected to be very uncertain, at least if it was performed in standardless ways.

It is to be noted that peak-to-background (P/B) based quantitative models [11] are independent from the detector efficiency per definition (for element X-ray energies >1 keV). It can be an additional indication why it was in the past more reliable than net-count based fundamental parameters-based models, like ZAF [12], with standardless analysis. The individual detector efficiency property deviations and uncertainty due to lamella grid structure disappear almost completely with the P/B-based measurement, also the question of alignment between detector and real X-ray emission point on the specimen surface is much less influenced, a point which was and is obviously still an issue with polymer windows grid lamellas. The uncertainty of efficiency stability and knowledge has even increased in the past with the trend to make detectors areas larger and to move them closer to the specimen X-ray emission point. The new silicon nitride window design has clear advantages in all points for the future, it reduces detector efficiency uncertainty and variation, and therefore it seems to be a significant improvement for the reliability of quantitative results with EDS technology with electron probe excitation, the EDS-EPMA.

6. Conclusions

There was an investigation and revision of the well-known transmission with X-rays for the Moxtek polymer window AP3 for measurements with EDS in the SEM. The reason for the required revision was the discovered transmission effects at Moxtek windows lamella edges, assumed to be mainly due to the divergent X-ray beam in SEM measurements with EDS. It was tried to separate the detector effects with absorption in X-ray windows of the detector completely from the inner-specimen excitation and absorption processes, by using two detectors with different windows in simultaneous measurements. The quantitative results become independent from the detector window, which was used for the measurements, after using the revised view about polymer window lamella influence in SEM measurements with EDS. Otherwise, it was not possible to use the more recent silicon nitride window for reliable standardless quantitative analysis, at least not with same model and parameter-sets. The results have also shown that the detector efficiency is not possible to get stable and reliable determined for L- and M- X-ray line series below an energy of about 600 eV. It is an additional pitfall for low-kV analysis techniques in general, at least if it is standardless performed. The recently introduced silicon nitride window (Si_3N_4, Amptek C2) is in general the better choice because the window-grid design avoids all possible extra-geometry effects for EPMA analysis with EDS. It will most likely contribute to further reduce the uncertainty within the EDS-EPMA technology with EDS-detectors.

References

[1] Alvisi M *et al.* 2006 *Microsc. Microanal.* **12** 406
[2] MOXTEK Data Sheet WIN-DATA-1006, Rev D
[3] Törmä P T, *et al.* 2013 *IEEE Trans. Nucl. Sci.* **60** 1311
[4] Törmä P T, *et al.* 2014 *IEEE Trans. Nucl. Sci.* **61** 695
[5] Amptek "Patented C-series low energy X-ray windows". www.amptek.com
[6] Eggert F, *et al.* 1991 *Rev. Sci. Instrum.* **62** 741
[7] Camus P P 2017 *Microsc. Microanal.* **23** 1064
[8] Rafaelsen J, *et al.* 2015 *Microsc. Microanal.* **21** 1645
[9] Eggert F, et al. 2021 *Microsc. Microanal.* **27** 1674
[10] Eggert F 2022 https://doi.org/10.13140/RG.2.2.33494.11849/1
[11] Eggert F 2018 *Microsc. Microanal.* **24** 734
[12] Eggert F 2020 *Microscopy Today* **28** 34

Machine Learning techniques in microscopic characterisation of nanomaterials

Benedykt R. Jany

Jagiellonian University, Marian Smoluchowski Institute of Physics, Faculty of Physics, Astronomy and Applied Computer Science, Lojasiewicza 11, 30348 Krakow, Poland

e-mail: benedykt.jany@uj.edu.pl

Abstract. The characterisation of nanomaterials presents significant challenges due to their unique properties and size-dependent behaviours. Recent breakthroughs in Machine Learning (ML) techniques have enabled the development of innovative methods for analysing microscopic data, thereby facilitating the extraction of reliable information on nanomaterials' chemical composition, structure, and other properties at the nanoscale level. This study provides an overview of selected ML-based approaches applied to microscopic characterisation of nanomaterials, including chemical composition quantification, structure analysis via imaging, and nano-electrical characterisation. The results demonstrate that ML techniques can significantly enhance and simplify the task of nanomaterials characterisation, while minimising data analysis bias and uncertainty. This approach has the potential to derive a comprehensive understanding of a given material's properties by integrating information obtained from diverse characterisation techniques. The synergistic coupling between microscopic imaging and Machine Learning techniques provides new quality, perspectives, and opportunities for future materials characterisation.

1. Introduction
Nowadays variety of different types on nano-materials are used all around us in different fields of science, technology and industry. Because of their size in single or tens of nanometres the appropriately matched techniques are used for their visualisation and later also characterisation. Microscopy techniques like transmission or scanning electron microscopy (TEM, SEM) together with atomic force microscopy (AFM) are commonly used. Due to the digital acquisition and storage big amount of data are collected by these techniques in the form of microscopy images or hyperspectral data from different type of spectroscopy measurements at nano-scale, etc. It is usually a challenging and tedious task to get reliable information on nano-materials characterisation such as quantitative chemical composition at the nano-scale, structure changes, and various other properties, from all the collected data. In the following, the applications of the selected Machine Learning (ML) techniques, which improve and simplify the task of characterisation of nano-materials via chosen microscopy techniques will be shown.

2. Machine Learning supported SEM EDS chemical quantification at nanoscale
Energy-dispersive X-ray spectrometry in the SEM (SEM-EDS), together with Machine Learning data processing could be used to quantify chemical composition at the nanoscale [1]. Systems such as $AuIn_2$ metal alloy nano-wires on an InSb substrate (AIIIBV semiconductor) could be successfully quantified.

The $AuIn_2$ nano-wires were synthesised in the processes of thermally induces self-assembly of 2 monolayers of gold on atomically flat InSb(001) surface at 330 °C in ultra-high vacuum conditions (UHV) [2]. The formed nano-wires are of an average width of ~ 70 nm. The first step towards the quantification is to collect the EDS data at the nano-scale from single nano-wires. The measurement condition (beam current and beam energy) were carefully optimised and verified by CASINO [3] Monte Carlo simulation of energy loss of electrons together with X-ray radial distributions for Au, In and Sb at 6.5 keV electron beam, see Figure 1 (top). The SEM-EDS data were acquired at 6.5 keV beam energy and 16 nA beam current, using double beam SEM/FIB FEI Quanta 3D FEG microscope equipped with EDAX Ametek Apollo XPP SDD EDS detector with an active area of 10 mm^2. The EDS data were acquired in the hyperspectral mode (EDS spectrum image (SI)), i.e., for each x- and y-position a full EDS spectrum was recorded, with a step size of ~29 nm. The collected raw data are available in [4].

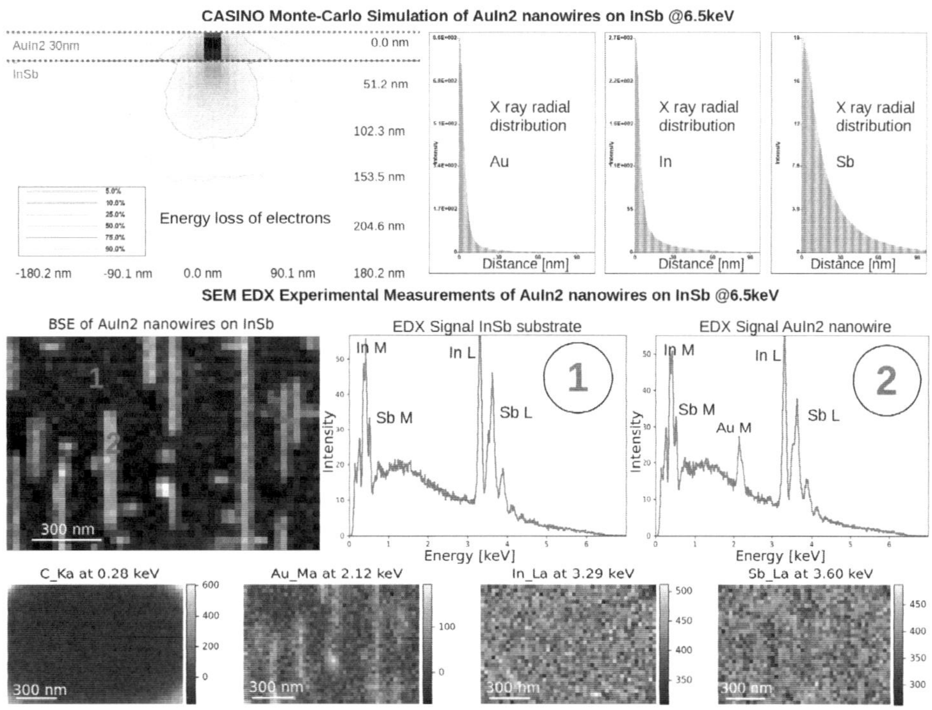

Figure 1. $AuIn_2$ nano-wires on an InSb surface. Top: CASINO Monte Carlo simulation results for 6.5 keV electron beam show energy loss of electrons and X-ray radial distributions for Au, In, and Sb elements. The X-ray signal originates from a depth of approximately 200 nm. Bottom: Backscattered electron image along with corresponding X-ray maps of C, Au, In, and Sb elements. SEM-EDS spectra are displayed for two regions: 1) EDS signal from the InSb substrate region, 2) EDS signal from the $AuIn_2$ nano-wire region. Due to the X-ray generation volume, the EDS signal from the nano-wire region contains not only the signal from the nano-wires but also the signal from below (i.e., InSb substrate).

Figure 1 (bottom) shows analysis results of collected SEM-EDS at the nanoscale [4]. The Au-Mα X-ray map shows clearly the nano-wire structures, similar to the simultaneously collected backscattered electron (BSE) image. The EDS spectrum from the $AuIn_2$ nano-wire region (Figure 1 bottom (2)) in addition to the Au and In X-ray signals (coming from $AuIn_2$ nanowire) shows also Sb-signal coming from the InSb substrate. This is related to the X-ray generation volume: The EDS signal from the nano-wire region contains also the X-ray signal from below, i.e., from the InSb substrate. The EDS signal from the $AuIn_2$ nano-wires is mixed with a signal from the InSb substrate. In the X-ray spectrum there is also a C-Kα signal, which is related to the carbon capping layer and contamination. Next, the collected SI EDS data [4], consisting of 1,536 EDS spectra, were processed by Machine Learning using the open-source software HYPERSPY [5]. First, principal component analysis (PCA) was performed to determine the number of components/phases present in the data. Three components were found to have significantly higher variance, as can be seen on the PCA scree plot (Figure 2). In the next step, non-negative matrix factorisation (NMF), with non-negative double singular value decomposition initialisation, was used to un-mix the data assuming three components found by PCA.

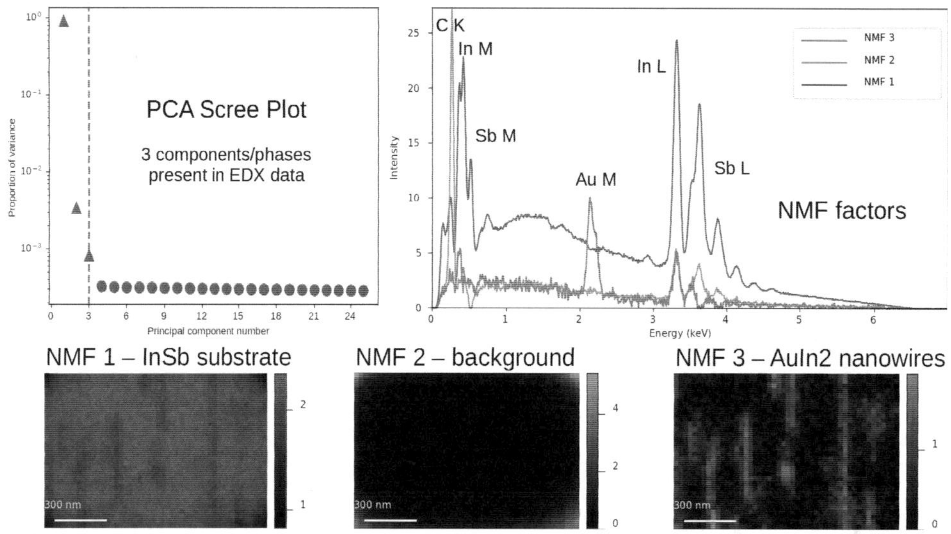

Figure 2. Results of Machine Learning processing via NMF and PCA on SEM-EDS SI data from $AuIn_2$ nano-wires on an InSb substrate. The PCA scree plot indicates three components/phases present in the data. Non-negative matrix factorisation (NMF) component maps and corresponding spectra reveal X-ray lines of elements: NMF 1 – InSb substrate, NMF 2 – background, and NMF 3 – $AuIn_2$ nano-wires. This demonstrates that NMF successfully separates the EDS signal from the nano-wires.

The NMF assumes the data to be non-negative. The results of the NMF decomposition of the SI EDS data is presented in Figure 2. NMF component maps and corresponding component spectra containing elemental X-ray lines show the following: NMF 1 – InSb substrate, NMF 2 – background (carbon and possibly secondary fluorescence), and NMF 3 – $AuIn_2$ nano-wires. It is worth to mention that the used NMF decomposition successfully separated the EDS signal coming from the nano-wires from other EDS signal sources, in particular successfully separation was performed between In and Sb, for the complicated structure of InL and SbL, which contain many lines.

Subsequently, in order to determine the quantitative chemical composition of the phases extracted by NMF (NMF 1 – InSb substrate, NMF 3 – $AuIn_2$ nano-wires), EDS ZAF standardless quantification was performed using EDAX GENESIS software. The results are presented in Table 1. This shows that the chemical composition determined by NMF decomposition together with ZAF quantification is very similar to the expected composition within the estimated uncertainties. The obtained results were further validated by detailed Monte Carlo simulations using DTSA2 software [6] from NIST, as well as additional TEM-EDS cross-sectional measurements [1].

Table 1. Results of EDS ZAF standardless quantification of NMF decomposition spectra together with expected composition derived from stoichiometry.

Decomposition component	EDS ZAF quantification (at%)	Expected composition (at%)
NMF 1 – InSb substrate	In: 45.0 (4.7) Sb: 55.0 (5.7)	In: 50 Sb: 50
NMF 3 – $AuIn_2$ nanowires	Au: 38.0 (5.9) In: 62.0 (5.9)	Au: 33.33 In: 66.67

The presented Machine Learning processing approach via NMF of the SEM-EDS SI maps collected at the nanoscale leads to the successful extraction of EDS signal coming from nanostructures, on which later chemical quantification is performed via EDS ZAF method. The spatial limitation of SEM-EDS, related to the X-ray generation volume effects, is greatly improved. The quantitative chemical composition of nanostructures could be successfully obtained by using only simple and achievable SEM-EDS measurements, which could lead to speeding up the measurements and reducing costs in comparison to costly and time-consuming TEM measurements [1]. The presented approach could be used to chemically quantify many different systems at nanoscale by SEM-EDS.

3. Machine learning based automatic microscopic image analysis

The machine learning approach could be applied to fully automatic microscopic image analysis [7, 8], see Figure 3. In the presented approach, a moving local window with a defined size is translated in the X- and Y-directions across the microscopic image. The window size (local region of interest on the image) defines the sensitivity scale of the analyses, a 128 pixel size seems to be optimal for different microscopic images of a size of ~ 2,000 × 2,000 pixels, producing the highest number of PCA components, for details please see [7]. In the next step, for each local window a 2D power spectrum Fourier-transform is computed. Thus, after this the single image data now consists of two spatial dimensions X and Y and correspondingly two dimensions in reciprocal space KX and KY, this together forms 4D data, very similar to the scanning precession diffraction data (SPED) 4D data. Next, this 4D data is processed by Machine Learning methods. To determine the components present in the data, the PCA scree plot is calculated, as implemented in HYPERSPY [5] and SCIKIT-LEARN [11]. The programme automatically analyses the PCA scree plot and determines automatically the number of components, which is later used for data decomposition using NMF [5, 11], with non-negative double singular value decomposition initialisation. This is done fully automatic without any user input. The output of NMF decomposition, similarly as for the SEM-EDS case of Figure 2, consists of interpretable spatial maps (loadings) and corresponding 2D power spectra (factors). The image is automatically analysed, i.e., the local features are discovered based of the local FFT, which is sensitive to local orientation and spacing.

This automatic approach was successfully applied to analysis of various microscopic images with and without local periodicity, i.e., SEM, atomically resolved high angle annular dark field (HAADF), scanning transmission electron microscopy (STEM), scanning tunnelling microscopy (STM), and fluorescence microscopy [7, 8]. It could also be successfully applied to other than microscopic images such as photography or a painting, to analyse local features on them [7].

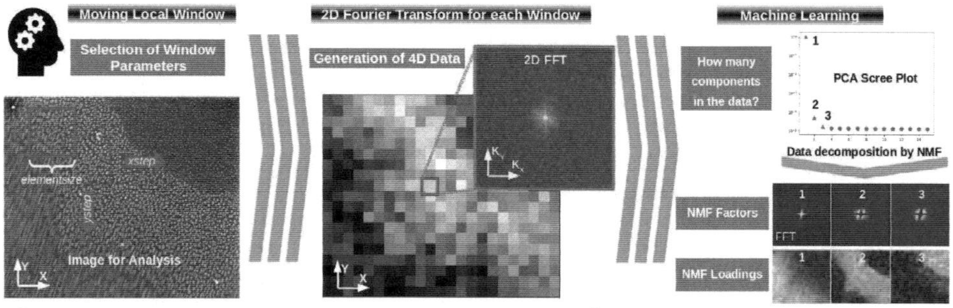

Figure 3. Automatic image analysis through Machine Learning and Fourier transform. A moving local window is applied to a microscopic image. For each local window, a 2D power spectrum Fourier-transform is computed. The data form a 4D dataset which is then processed by machine-learning: 1) PCA scree plot to determine the number of components/fractions in the data. 2) NMF decomposition the data into interpretable components. Local features are automatically discovered within the image through this process. (From [7] under a Creative Commons license CC by 4.0).

The fully automatic analysis of a single image takes about a minute on an average desktop or notebook computer. This gives the possibility to analyse, within a reasonable time and with computing resources, large amounts of image data taken during measurements on the microscope. The programme outputs a PDF file with a report from analysis. The analysis approach is freely available as a Python analysis notebook and Python programme for batch processing from Mendeley Data [8] (https://dx.doi.org/10.17632/25x46xjyr5.3).

For example, Figure 4 present the use of the automatic approach to a SEM image (1,662 × 1,370 pixels) of Au-rich nano-structures on a GaSb(001) surface shows three regions with three different types of nano-structures.

These regions are the consequence of different surface stoichiometry related to the non-uniform enrichment of the clean surface in gallium. The analysis, which takes about a minute, automatically finds these three regions in the data (NMF loadings and corresponding NMF factors). From the NMF factors, which are the 2D power spectra characterising each region, one can calculate the characteristic spacing between the nano-structures and the directionality of the structures.

Another example is the application of the automated programme approach to the analysis of an atomically-resolved HAADF STEM image of Au nano-islands grown on a reconstructed Ge(001) surface (Figure 5). The formed nano-islands are made of two phases of Au, i.e., a cubic fcc-phase and a rare and unusual hexagonal hcp-phase, separated from each other by a defected inter-growth domain along (111) crystallographic planes [9]. The atomically resolved image shows the germanium bulk at the bottom, the nano-island in the middle, and the platinum capping layer on the top. It is not trivial to the naked eye to distinguish in the nano-island the regions of the two different Au phases, i.e., the Au fcc and Au hcp. The analysis, which takes about a minute, automatically finds the regions of the different Au phases. The NMF loadings shows clearly how these regions are located and how they merge into each other. The corresponding NMF factors 2D power spectra characterise each region corresponding to a different phase.

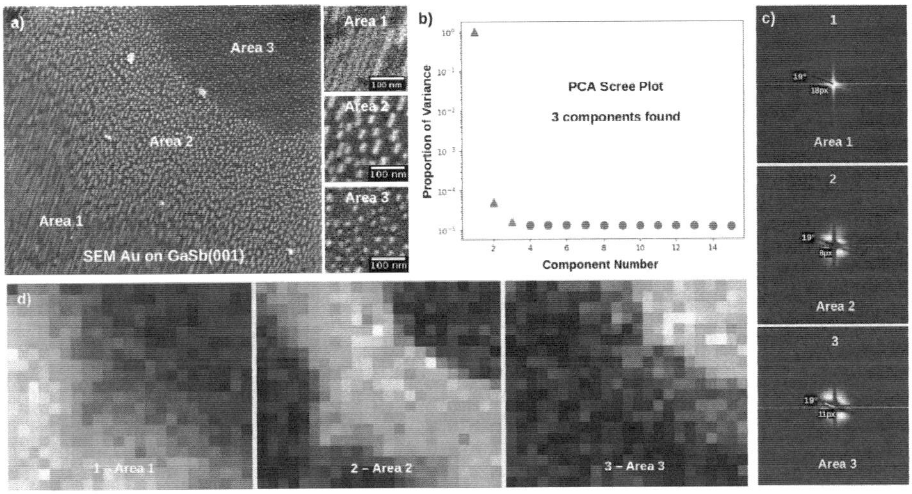

Figure 4. Results of the approach on a SEM image of Au-rich nanostructures on GaSb(001) surface. a) The image shows three different morphologies: Nano-wires, elongated islands, or symmetric clusters. b) PCA scree plot highlights variance distribution across the first 15 principal components, with red marking the top three. c) and d) NMF decomposition results of moving local window Fourier-transform data are displayed as factors and loadings (Fourier transforms). The approach reveals details about the nanostructures morphology. Image width is 3 μm. (From [7] under a Creative Commons license CC by 4.0).

4. Analysis of conductive AFM current-voltage (I-V) data by Machine Learning

Conductive AFM (C-AFM) is a common technique of local electrical properties characterisation of nano-scale electronic devices like nano-electrodes. Usually the electrical properties are measured by the conductive AFM tip by acquiring in each image point a full current-voltage (I-V) curve, which forms a hyperspectral data cube. One typical C-AFM measurement in this mode consists of several thousands of I-V curves, which need to be analysed and interpreted. Here, the Machine Learning methods could be also used to simplify and to enhance the electronic nano-device characterisation [10]. The metallic nano-electrodes grown on an n-doped InP(001) surface were successfully electrically characterised by C-AFM together with Machine Learning processing [10]. The formed nano-electrodes on InP exhibit two types of morphologies, i.e., "Flat Top" and "Sharp Top" as seen in SEM and AFM topography, see details in Figures 6a to 6d. These two different types of morphologies are a consequence of two different types of crystallographic orientation of the metallic nano-electrodes and different atomic structure at the nano-electrode substrate interface, as seen in atomically-resolved HAADF STEM [10]. In consequence, these influences the electronic properties of the nano-electrodes, which were characterised by in situ C-AFM measurements in hyperspectral mode, see Figures 6e and 6h. More than 7,000 I-V curves were collected. The Machine Learning method *k-means* clustering, as implemented in SCIKIT-LEARN [11], was used to group the collected I-V data into clusters with similar properties. This is accomplished by iteratively minimising the sum of squares between the points and cluster centres (in the multi-dimensional space) defined as the arithmetic mean of all points, which belong to the cluster. The C-AFM I-V data were clustered into three groups presented as a map showing different I-V regions, together with corresponding average I-V characteristics in these regions.

Three different regions are visible: Low conductivity nano-electrodes region (blue), which exhibit non-linear I-V behaviour, high conductivity nano-electrodes region (red) with linear I-V characteristic, and an InP surface region (green). It is seen that ~ 70 % of all nano-electrodes are of lower conductivity. The used Machine Learning clustering significantly improved the analysis and interpretation of C-AFM I-V measurements [10].

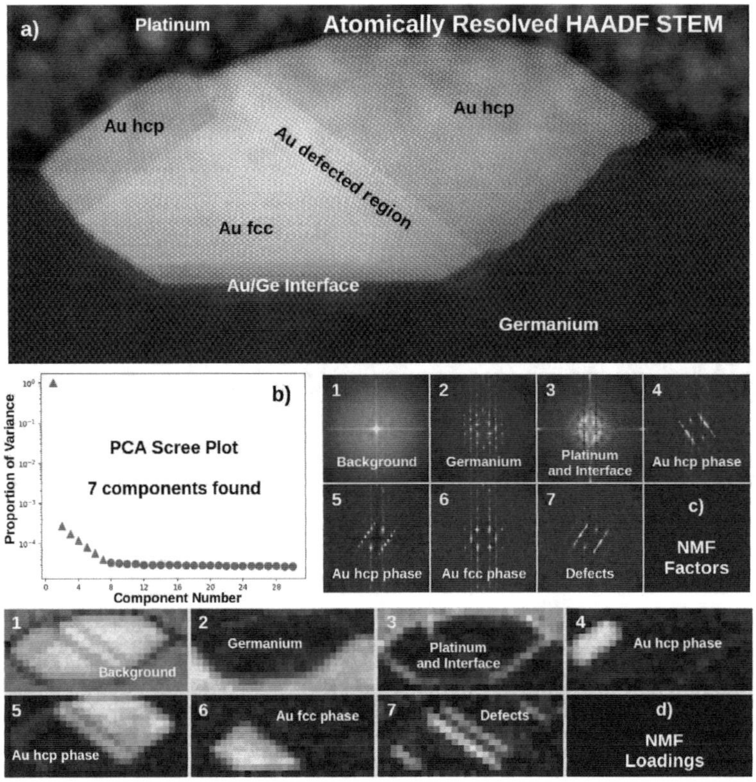

Figure 5. Results of the approach on an atomically-resolved HAADF STEM image. a) A nano-island with Au fcc/hcp phases grown on Ge(001) [9] (Creative Commons license 4.0). b) PCA scree plot shows variance distribution across the first 30 principal components, with significant differences in the top seven. c) and d) NMF decomposition results of moving local window Fourier-transform data are displayed as factors and loadings. The approach reveals details about the nano-island structure. Image width is 49.7 nanometres. (From [7] under a Creative Commons license CC by 4.0).

5. Conclusions

The presented selected Machine Learning based approaches to microscopic data greatly enhance and simplify the task of nanomaterials characterisation in the areas of chemical composition quantification [1], structure analysis via microscopic imaging [7], and nano-electrical characterisation [10]. The usage of Machine Learning techniques to microscopic characterisation of nanomaterials allows for reliable extraction of materials properties at the nanoscale level. This kind of approaches are very beneficial and attractive way of data analysis to obtain reliable output from measurements minimising the analysis bias, i.e., keeping the same bias for all the data (computer algorithms bias). The limitations of this type of approaches are related to the input data quality. For these methods to work, experimental data of reasonable high quality are needed, i.e., the collected data must have an appropriate statistics (for hyperspectral data it is good to have around ~100 counts in ROI window). Experimental artefacts like different kind of noise can corrupt the data in the way that effectively the proper operation of ML methods will be blocked. Before proceeding with Machine Learning techniques the data must be properly prepared so the experimental artefacts will be suppressed by the techniques of data curation (data cleaning, etc.). It is up to the user (microscopist) to provide high quality data from experiments to ML approaches to obtain reliable results.

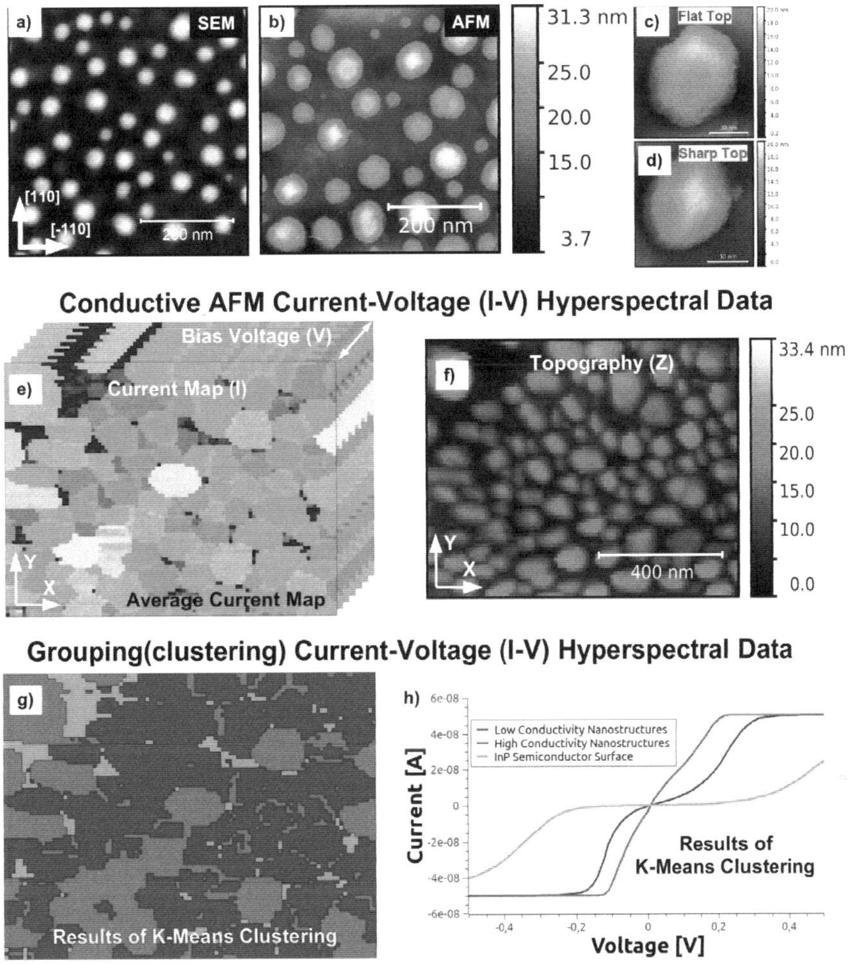

Figure 6. Morphology and electrical properties of $AuIn_2$ nanostructures on n-doped InP(0 0 1) surface. a) SEM image shows the surface morphology. b) AFM topography reveals details about the nanostructures. c) and d) Two types of nano-electrodes are visible: "Flat Top" and "Sharp Top". Flat Top has low conductivity, while Sharp Top exhibits high conductivity. e) C-AFM average current map highlights differences in electrode behaviour, together with f) corresponding topography. g) C-AFM I-V hyperspectral data clustering reveals three regions: low conductivity (blue), high conductivity (red), and InP surface region (green). h) Average I-V characteristics for each region are shown, with 70 % of nano-electrodes being lower conductivity. (From [10] under a Creative Commons license CC by 4.0).

The Machine Learning methods are not limited only to the presented examples, they could be also successfully used to automate and improve different tasks of materials characterisation in diverse fields of microscopy such as in electron backscatter diffraction (EBSD) data analysis or HAADF image quantification [2]. The Machine Learning methods are also suitable to light microscopy techniques like the colorimetric microscopy (C-Microscopy) approach [12, 13].

There is also a great potential for advancing our understanding of material properties, improving efficiency and enabling new discoveries based on the Machine Learning techniques in microscopy for materials characterisation. These tools could be applied in the fields of automated image analysis, material design and optimisation, materials defects detection and characterisation, material classification based on microscopic properties, automation of sample preparation and handling. There exists also a requirement for employing Machine Learning strategies in the acquisition of material characteristics from intertwined data gathered through various microscopic, spectroscopic, and diffraction methods. This approach kind of could potentially enable derivation of a comprehensive understanding of a given material attributes by integrating information obtained from diverse techniques.

The synergetic coupling between microscopic and Machine Learning techniques provides new quality and perspectives for the future in materials characterisation.

Acknowledgements
This research was supported in part by the Excellence Initiative - Research University Programme at the Jagiellonian University in Krakow.

References

[1] Jany B R, Janas A and Krok F 2017 *Nano Lett.* **17** 6507-7170
[2] Jany B R, Janas A, Piskorz W, Szajna K, Kryshtal A P, Cempura G, Indyka P, Kruk A, Czyrska-Filemonowicz A and Krok F 2020 *Nanoscale* **12** 9067-9081
[3] Drouin D, Couture A R, Joly D, Tastet X, Aimez V and Gauvin R 2007 *Scanning* **29** 92
[4] Jany B R, Janas A and Krok F 2017 Collected Au/InSb and Au/Ge SEM EDX experimental data at nanoscale as used in retrieving the quantitative chemical information at nanoscale from scanning electron microscope energy dispersive X-ray measurements by machine learning. http://dx.doi.org/10.13140/RG.2.2.19558.93763
[5] de la Peña F, Ostasevicius T, Fauske V T, Burdet P, Jokubauskas P, Nord M, Sarahan M, Johnstone D N, Prestat E, Taillon J, Caron J, Furnival T, MacArthur K E, Eljarrat A, Mazzucco S, Migunov V, Aarholt T, Walls M, Winkler F, Martineau B, Donval G, Hoglund E R, Zagonel L F, Garmannslund A, Gohlke C and Chang H-W 2017 http://doi.org/10.5281/zenodo.345099
[6] Ritchie N W M 2011 *Microscopy Today* **19** 26-31
[7] Jany B R, Janas A and Krok F 2020 *Micron* **130** 102800
[8] Jany B R 2020 *Python Jupyter notebook to perform automatic microscopic image analysis by moving window local Fourier transform and machine learning. Mendeley Data, v3.* https://dx.doi.org/10.17632/25x46xjyr5.3
[9] Jany B R, Gauquelin N, Willhammar T, Nikiel M, van den Bos K H W, Janas A, Szajna K, Verbeeck J, Van Aert S, Van Tendeloo G and Krok F 2017 *Sci. Rep.* **7** 42420
[10] Janas A, Piskorz W, Kryshtal A, Cempura G, Belza W, Kruk A, Jany B R and Krok F 2021 *Appl. Surf. Sci.* **570** 150958
[11] Pedregosa F, Varoquaux G, Gramfort A, Michel V, Thirion B, Grisel O, Blondel M, Prettenhofer P, Weiss R, Dubourg V, Vanderplas J, Passos A, Cournapeau D, Brucher M, Perrot M and Duchesnay E 2011 *J. Mach. Learn. Res.* **12** 2825-2830
[12] Jany B R 2024 *Micron* **176** 103557
[13] Jany B R 2023 *Python Jupyter Notebooks and Data for Colorimetric Microscopy (C-Microscopy) Approach. Zenodo.* https://doi.org/10.5281/zenodo.7789585

IOP Conf. Series: Materials Science and Engineering **1324** (2025) 012008 doi:10.1088/1757-899X/1324/1/012008

100 years of X-ray Kossel micro-diffraction and its advantages versus the exclusive application of EBSD: From the prediction to current developments
A tribute to the memory of Priv.-Doz. Dr. habil. Siegfried Däbritz

E Langer

Dresden University of Technology, Institute for Semiconductors and Microsystems, 01062 Dresden, Germany

E-mail: langer@physik.tu-dresden.de

Abstract. The year 2024 marks the 100th anniversary of the prediction of X-ray interferences from lattice sources by W. Kossel. The μ-X-ray Kossel Diffraction (μXKD) can be easily applied in combination with EBSD as additional complementary diffraction method in a SEM or in a microprobe. It has the highest informational content and precision compared to conventional diffraction methods. It is simply possible to use one single scintillator CCD detector combination to observe both diffraction methods, which will be especially discussed more in detail here, as our group has successfully demonstrated over 25 years. A historical overview of selected applications will be given as review, besides of precise local lattice constants measurements, the determination of phase angles of the structure factor and polar planes in non-centrosymmetric crystals at the semiconductors GaP and GaAs, the estimation of dislocation densities in sinter model materials, the local measurement of the thermal conductivity as well as thermal expansion coefficients and the phase identification in Au-Ti diffusion zones as example.

1. Introduction

1.1. The 100th anniversary of the prediction of the interference of the excited characteristic X-rays in a crystal itself at the Kiel University
At the discovery of the diffraction of X-rays by crystals by von Laue *et al.* in 1912 initially the interference of fluorescent radiation excited in the crystal had been erroneously supposed to be the cause of the diffraction of X-rays by crystals. M. von Laue recognised quickly that the direct diffraction of the incident X-rays is the right interpretation. This idea of the diffraction of the secondary radiation was rejected again and was not recognised as a new interference method until Kossel predicted in 1924 the interference of characteristic X-rays in the crystal, in which the self-radiation is excited, the so-called X-ray interferences from lattice sources [1]. Ten years later this X-ray interference effect was proven by Kossel *et al.* in 1934 [2]. Figure 1 shows a photograph of Walther Kossel.

Figure 1. Walther Kossel, 1888 - 1956 (from the Kossel-Möllenstedt library, Dresden)

1.2. Principle of the lattice source X-ray interferences or X-ray Kossel diffraction

The Kossel effect exploits the emitted characteristic X-rays within the investigated crystalline volume. Thereby, the excitation is mainly performed with electrons [2, 3], but also with X-rays [4], protons [5], ions [6], or synchrotron X-ray [7, 8], respectively. Later the X-ray excitation of Kossel patterns with a focussing polycapillary lens using an X-ray fluorescence (XRF) spectrometer [9] and with a compact X-ray source at a SEM was shown [10]. In the near future, the excitation by an X-ray free-electron laser is conceivable [11]. In directions satisfying Bragg law rotationally symmetric sharp cones appear, the so-called Kossel cones, named to Kossel's honour, see Figure 2. These complex interference phenomena, which have to be dealt by the dynamical theory of X-ray diffraction [12], nowadays, appear on a planar CCD detector [13-16, 18] as pure dark and bright Kossel lines and such one with a bright-dark fine structure (see line for example the profile of the 220 reflection in Figure 4). The characteristic X-rays of the chemical elements are excited for the Kossel effect, for example at an iron monocrystal only the Fe-K$\alpha_{1/2}$-radiation with high intensity and this then interferes, see Figure 9 on the right side.

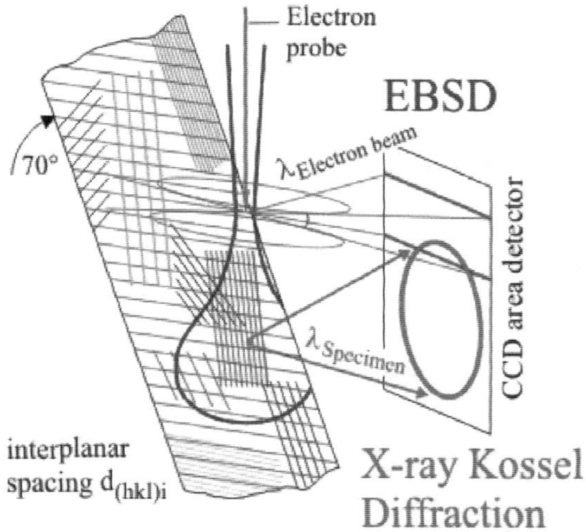

Figure 2. Basic principle of the Kossel effect with electron excitation in combination with EBSD. The Kossel curves are conic section: circles, ellipses, parabolas, and hyperbolas, but only one Kossel half-cone with the corresponding circle reflection is shown and marked in red colour. Additionally a pair of Kikuchi lines can be seen, which is coloured blue.

1.3. Motivation to apply additional the X-ray Kossel μ-diffraction versus the exclusive application of EBSD

It has become apparent that electron backscatter diffraction (EBSD) is an advantageous method with high lateral resolution and fast mapping speed for phase discrimination, orientation determination and texture analysis as well as is characterised by an increasing relative accuracy through the development of the cross correlation method by Wilkinson *et al.* [19, 20]. Recently, the intensity simulation by the dynamical diffraction theory by Winkelmann *et al.* [21, 22] led to further crystal information, like the point-group sensitive orientation mapping on non-centrosymmetric phases by Nolze *et al.* [23], see Table 1. However, the X-ray Kossel diffraction has a high informational content, and provides information also on the crystal symmetry and crystal structure, superstructures, exact orientations and misorientations, on dislocation densities, allows the precise determination of local lattice constants up to a relative accuracy of 10^{-5}, of tetragonal distortions with a shown accuracy of $\Delta(c/a) = \pm 0.0006$ [24] as well as residual stresses all three in the μ-range. For the last-mentioned parameters, the attainable precision is unreachable with EBSD, see Tables 1 and 2. Therefore, the Kossel technique is an extremely valuable additional method in combination with EBSD. Typical measurements can start with an EBSD mapping and in a second step some Kossel patterns on problematic (e.g., non-identified black pixel regions) or other selected measuring points will provide missing or more precise additional crystal information. Table 1 summarises the most important crystal information, which can be determined by X-ray Kossel diffraction compared to EBSD.

Table 1. Comparison of parameters and applications of Kossel method compared with EBSD.

Crystal parameters and applications (current state of research in 2024)	Kossel	EBSD
· Automated crystal orientation mapping and texture analysis		● ● ●
· High speed of acquisition of the diffraction patterns		+++
· Greatest lateral resolution of the diffraction patterns		+++
· Automated pattern quality mapping		+++
· Automated phase discrimination map		++
· Polarity-sensitive orientation mapping on non-centrosymmetric phases		++
· Polar plane distinction on non-centrosymmetric crystals	+++	++
· Phase identification	++	+
· Crystal structure determination	++	+
· Symmetry determination, e.g. tetragonal distortion	+++	+
· New formation of phases and phase change in the high- and	++	+
low temperature range	++	+
· Exact determination of crystal orientations: $\pm$ 0.01°	+++	++
· Proof of mechanical tensions and deformations	+++	+
· Residual stress measurements	+++	+
· Precise lattice constant determination $\Delta a/a \approx 10^{-5}$	● ● ●	)[1]
· Phase determination from the structure factor	++	
· Direct estimation of dislocation densities	++	
· Determination of expansion coefficients	+++	
· Indirect determination of chemical concentration	++	
· Detectability, location of crystal defects	+	

● ● ● main application; +++: excellent; ++: very good; +: possible. Updated from [16, 17].

[1] It is possible to estimate lattice constants by EBSD with an accuracy in the order of $\Delta a/a \approx 10^{-2}$, e.g. [25].

Table 2. Comparison of the precision of both diffraction methods.

	Kossel Diffraction	Electron Backscatter Diffraction
● Symmetry determination, e.g. tetragonal distortion		
Precision:	$\Delta(c/a) = \pm\ 0.0006$	$\Delta(c/a) = \pm\ 0.01$
Lowest detectable deviation	high precision of c/a = 1.002	advantage of higher spatial resolution c/a = 1.01
Evaluation method:	multiple line intersections	by simulated reference patterns
Reference:	Ullrich et al. [24]	Nolze et al. [25]
● Precise lattice constant determination		
Precision:	$\Delta a/a \approx 10^{-5}$ in special cases $3 \cdot 10^{-6}$	$\Delta a/a \approx 10^{-2}$ method is too imprecise
Reference:	Bauch et al. [26]	
● Crystal lattice orientation determination		
Absolute orientation		
Precision:	0.1°	1°
Evaluation method:	various methods	e.g. standard Hough transform
Relative orientation		
precision:	0.01°	(0.5 - 0.1)°
Evaluation method:	e.g. superposition of Kossel pattern and simulation	standard Hough transform 0.01° [18, 19]
Evaluation method:	further various methods	pattern cross-correlation based
References:	Bauch et al. [26]	Wilkinson et al. [19, 20]
● Residual stress measurements		
Detailed information	Allows absolute and relative residual stress measurements, that means distinguishable residual stress measurements of the 1st, 2nd and 3rd kind	Allows residual stress measurements relative to reference points in the same crystallite, that means residual stress measurements of the 3rd kind
Precision:	10^{-5}	10^{-4} [19, 20]
Evaluation method:	precision compensation method, Brechbühl et al. [27]	pattern cross-correlation based [19, 20]
Highlight:	Residual stress mappings [28]	
References:	Böhling et al. [28]	Wilkinson et al. [19, 20]

2. Great variety of applications for the X-ray Kossel diffraction in the SEM

These are a continuation of the paper by Langer and Däbritz [17], which did already treat the application examples precise lattice constant determination and tetragonal distortion, measurement of phase transformations at low (superconductor V_3Si) and high temperatures (high-temperature Kossel experiments on Cu and the α-γ transformation of iron), as well as the phase identification by means of the characterisation of intermetallic phases in the Cu-Sn system of Table 1. The applications of the X-ray Kossel diffraction summarised in Table 1, which are coloured in blue will be discussed here in this paper more in detail in the following sections.

2.1. Determination of the phase angle of the structure factor

For conventional X-ray diffraction methods in which the X-ray source lies outside of the crystal, by the experimental measured intensity $I_{hkl} \sim |\mathrm{F}(hkl)|^2$ of an X-ray reflection only the modulus of the structure factors $|F(hkl)|$ and not the phase angle α(hkl) of the structure factor F(hkl) can be evaluated (phase problem of the crystal structure analysis with X-rays or electron beams):

$$\mathrm{F}(hkl) = |\mathrm{F}(hkl)|e^{i\cdot\alpha(hkl)} \tag{1}$$

A direct solution lies in the Kossel lines, which show a remarkable bright-dark fine structure in the line profiles, their intensity was calculated by means of dynamical X-ray diffraction by von Laue [29, 30] in the year 1935. As early as in the year 1938 it was suggested by Ott [31] to apply these fine structure for the experimental direct determination of these scatter phase angles at the crystal structure analysis. Blau *et al.* [32] have discussed the concrete evidence concerning mainly the sign of the structure factor at experimental examples already in the year 1967. The determination of the phase information that means the phase angle especially at mixed structure factors by comparison between theoretical calculated and experimental Kossel line profiles considering the absorption were shown by Stephan *et al.* [33, 34] beginning on 1974 at many examples, like for instance Fe, ZnS, etc., but also as below for GaAs.

In the year 1975 Brümmer and Nieber [35] investigated the fine structure of the Kossel lines in dependence on the phase in the symmetric Laue case, see Figure 3, with the help of measured photometer curves, especially in GaP and GaAs at the reflections (220) and (111) Ga-Kα, see Figure 4. Kossel line fine structure and corresponding readable phase angles are listed in Table 3.

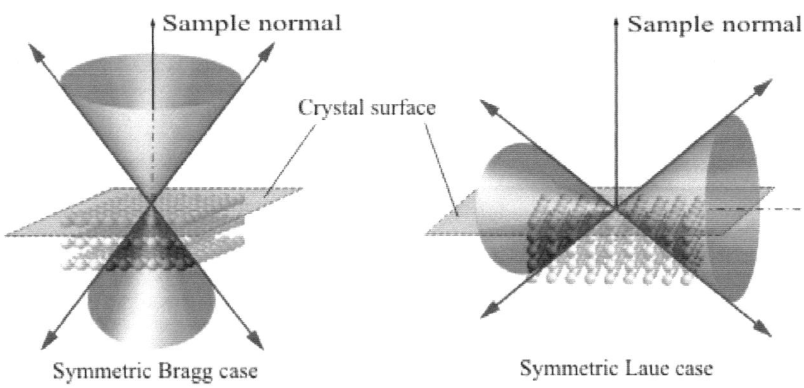

Figure 3. Special geometry cases of Kossel cones relative to the crystal surface.

10 years later Hutton *et al.* [36] investigated the feasibility to exploit Kossel lines in the Bragg case for the measurement of the phase angle. The line profiles of Ga (111) and Ga (333) Kossel reflections on GaAs were exemplary calculated for this purpose and, furthermore, they suggested the synchrotron excitation. Recently the experimental determination of the phases α(hkl) of the structure factor from measured line profiles on further Kossel reflections of GaAs succeeded quantitatively by synchrotron excitation in [37].

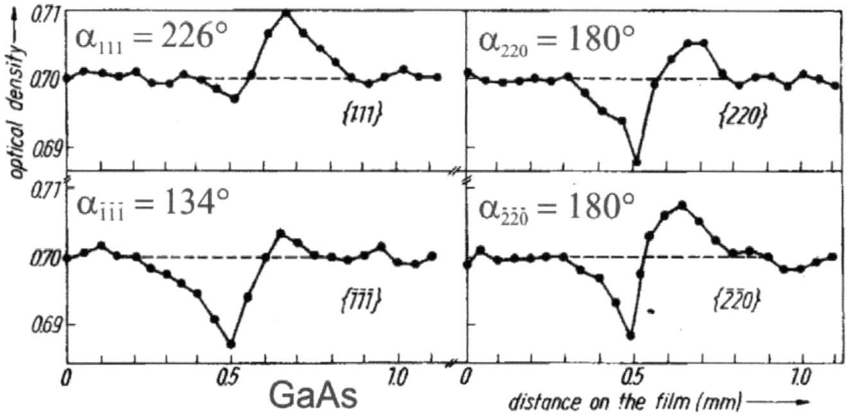

Figure 4. Two selected examples of photometer curves of the {1 1 1}, {1 $\bar{1}$ $\bar{1}$} and {2 2 0}, {$\bar{2}$ $\bar{2}$ 0} Ga-Kα type Kossel reflections in GaAs with the phase α(hkl) assigned to the reflections in each case inserted on the left side above from Brümmer and Nieber [35]. Higher optical density meant darker on a X-ray film.

Table 3. From the fine structure of the Kossel line profile simply readable qualitative phase angle: 0° ≤ phase angle α(hkl) ≤ 360° in the symmetric Laue case from [35].

Fine structure of the Kossel line profiles	Readable phase angle α(hkl)
Mainly dark Kossel line (see Fig. 4)	$\alpha(hkl) > 180°$
Mainly bright Kossel line (see Fig. 4)	$\alpha(hkl) < 180°$
Symmetric bright-dark Kossel line fine structure (see Fig. 4)	$\alpha(hkl) \cong 180°$

2.2. Polar plane distinction on non-centrosymmetric semiconductor crystals

Analysing the crystal structure symmetries and their determination play an important role. Conventional X-ray diffraction methods cannot prove an inversion centre, as expressed in the Friedel's rule, which states, that for X-ray diffraction the intensity of the reflections $h\ k\ l$ and $\bar{h}\ \bar{k}\ \bar{l}$ is identical. Also, for non-centrosymmetric crystals, i.e., do not have a centre of symmetry apply: $I_{\bar{h}\bar{k}\bar{l}} = I_{hkl}$.

Max von Laue's assumption of the Kossel effect on non-centrosymmetric crystals was, that it will be possible to distinguish between the directions $\langle h\ k\ l \rangle$ and $\langle \bar{h}\ \bar{k}\ \bar{l} \rangle$ by the Kossel effect, i.e., Friedel's law is violated here, which was verified by Brümmer *et al.* [35, 38] in the year 1973 for the first time.

The features in the interference phenomena, which appear in non-centrosymmetric compared to centrosymmetric crystals, are highly suitable for the determination of polar axes in non-centrosymmetric crystals by the Kossel effect and for an unambiguous distinction between non-centrosymmetric and centrosymmetric crystals [38, 35].

It can be seen particularly clear at the non-centrosymmetric III-V-semiconductor GaP of Figure 5 with zinc blende structure, since for P-Kα-radiation (λ = 0.6154 nm from [39]) only 8 reflections of the {1 1 1}-type occur. The bright Kossel circle emerges on the side of the small monocrystal plate with the Ga-atoms (A-face). The dark circle arises at the exposure of a Kossel pattern on the reverse side with the P-atoms (B-face) by Ullrich *et al.* in 1977 [40]. Further polarities were determined on the III-V compound semiconductors GaAs, GaP and InP for optoelectronic components and published by Nolze *et al.* in 1990 [41].

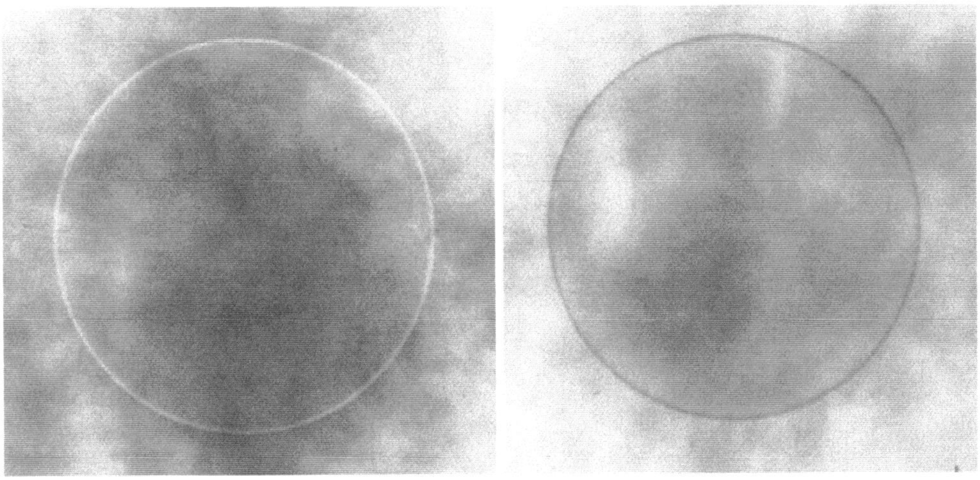

Figure 5. Kossel patterns of GaP of the year 1977. Left: the 111-P-Kα circle reflection on the (1 1 1). Right: the $\bar{1}\bar{1}\bar{1}$-P-Kα circle reflection on the $(\bar{1}\bar{1}\bar{1})$ crystal surface is clearly distinguishable [40].

2.3. *Direct estimation of dislocation densities: (10^6 - 10^{10}) cm^{-2}*

The line width of Kossel patterns clearly changes in the range of 10^6 to 10^{10} dislocations per cm^{-2} [42]. In order to estimate dislocation densities in the μ-range by the Kossel technique calibration samples of the same material with known different dislocation densities are required, to calibrate the correlation between dislocation density and line width. For measurement of the line width in the crystal volume to be investigated and on the calibration samples an identical point of the Kossel line in the same reflection system section must be utilised [42]. Because of the high symmetry such as for the cubic crystal structure this is easy to implement. Starting from a dislocation density of 10^9 cm^{-2} it was additionally necessary to adjust the same take-off angle of the measuring position on the reflection to the sample surface [42].

Measurements on reflections of the type {1 1 1} of monocrystalline Cu balls on (1 1 1)-oriented Cu single crystals in the sintering contact region especially at the $[\bar{1}\ \bar{1}\ \bar{2}]$-pole were carried out in the year 1983 by Ullrich *et al.* [42] and Herenz *et al.* [43], which are summarised in Figure 6. At dislocation densities between 10^9 and 10^{10} cm^{-2} a splitting of {111} Kossel lines were observed, whereby this appearance was discussed as typical for this range [42].

2.4. *Estimation of thermal conductivities in the micro-range*

The comparison of experimentally determined temperature increases by means of measurement the lattice constant changes for different absorption currents (constant accelerating voltage) with theoretical ascertained values, allows with known expansion coefficients α from literature the estimation of the thermal conductivity λ:

$$\text{with } \Delta T = \frac{\Delta a}{\alpha \cdot a_0} \text{ and } \Delta T = \frac{3 \cdot P}{2 \cdot d \cdot \pi \cdot \lambda}, \text{ it follows } \lambda = \frac{3 \cdot \alpha \cdot a_0 \cdot P}{2 \cdot d \cdot \pi \cdot \Delta a} \qquad (2)$$

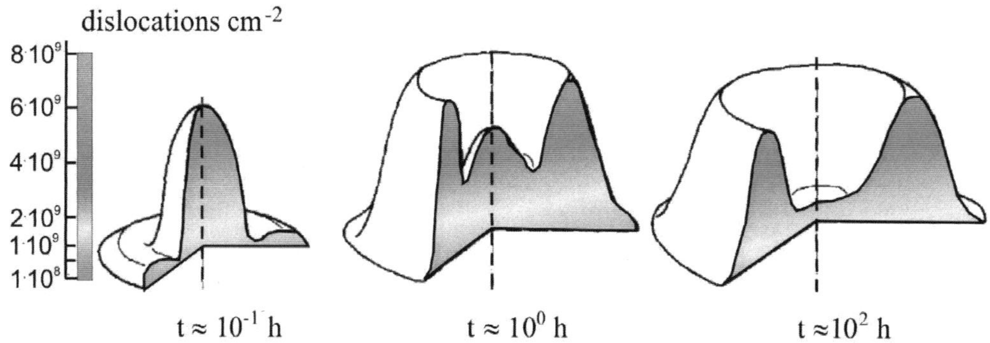

Figure 6. Idealised 3d representation of the measured dislocation density distribution by means of the Kossel diffraction from Ullrich et al. already in the year 1983 [42] depending on the different sintering times t at a sinter temperature of T = 900 °C.

where ΔT is the temperature increase, Δa is the lattice constant change between two different absorption currents measured by the Kossel technique, d is the diameter of the electron probe and P is the electrical power [44]. The lattice constant a_0 of the analysed area without a heating by the electron probe can be extrapolated for $i \to 0$ by the linear experimentally measured connection between lattice constant a and absorption current i. The conductivities λ of the compounds FeAl, $MgCu_2$ and the superconductor V_3Si were estimated already by Däbritz [44] in the year 1974 and are summarised in Table 4.

Table 4. Estimation of local thermal conductivities λ in the µ-range by Kossel diffraction by Däbritz [44]. For example in the superconductor V_3Si one can see the variation in different grains of the polycrystal.

Substance	U/kV	i/µA	d/µm	$\bar{T}$/K	$\alpha \cdot 10^{-6}/K^{-1}$	$\Delta a/(\alpha \cdot a_0)/K$	$\lambda/W \cdot mK^{-1}$
FeAl	60	2.8	35	293	18	81.3	28.2
FeAl	60	1.0	35	293	18	38	21.5
$MgCu_2$	60	1.0	35	293	31.5	3.6	230
V_3Si (grain 1)	50	2.3	50	100	4	205	5.4
V_3Si (grain 2)	50	2.7	50	100	4	230	5.6
V_3Si (grain 3)	50	4.0	50	100	4	347	5.5
V_3Si (grain 4)	50	2.0	50	170	4	159	6.0

2.5. Determination of thermal expansion coefficients

Vice versa, for a known thermal conductivity λ from the literature also a determination of the thermal expansion coefficients α is easily possible. In Table 5, the essential results of thermal expansion coefficient and lattice constant measurement by high-temperature Kossel investigations attained by Ullrich and Däbritz in the end of the 1960s are listed [44, 45, 46].

Table 5. Overview of high-temperature Kossel investigations regarding lattice constant changes for expansion coefficient measurements by Ullrich and Däbritz in the end of the 1960s [44, 45, 46]; α: thermal expansion coefficient ascertained, a: lattice constant determined.

Material	Temperature range/°C	Measured variables	Rel. precision of a determination
Fe	20 to 850	a(T), α(T)	$1{\cdot}10^{-4}$
Ni	20 to 110	a(T), α(T)	$3{\cdot}10^{-5}$
Cu	20 to 80	a(T), α(T)	$3{\cdot}10^{-5}$
Fe_3O_4	20 to 856	a(T), α(T)	$3{\cdot}10^{-4}$
Fe_3Al	20 to 100	a(T), α(T)	$3{\cdot}10^{-5}$
$MgCu_2$	20 to 80	a(T), α(T)	$3{\cdot}10^{-5}$

2.6. Phase identification by the Kossel diffraction with the example of Au-Ti diffusion zones.

The Au-Ti system was studied due to their significance for the hardness of gold jewellery [47]. Multi-phase diffusion zones in the Au-Ti alloys should serve as an example for the advantageous characterisation of a large variety of different crystal phases, which were investigated by the Kossel technique [48]. The presented sample was prepared by annealing a compact Au-Ti sandwich at 1,073 K for 216 h [48]. Figure 7 shows a SEM micrograph of the diffusion zone. The composition of each intermetallic compound has been measured by energy-dispersive X-ray spectrometry (EDS) in electron probe microanalysis (EPMA), whereas the plotted crystal phases were identified and the crystal orientations were determined by the Kossel diffraction [48]. In Figure 8, the sphere models of Au_4Ti, Au_2Ti and $AuTi_3$ can be seen, which are oriented in the direction of the measured crystal orientation in diffusion direction, whereby the special crystal orientation of the Au_4Ti crystal illustrates the similarity of the "open" directions to Au_2Ti, if one neglects the atom species [48]. The Kossel reflection systems were simulated with the self-developed programme KOPSKO 2.0 [49]. Table 6 gives an overview of the crystal data of the different phases.

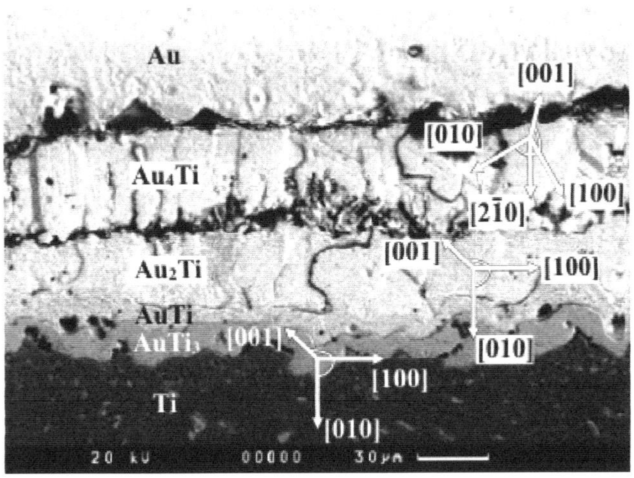

Figure 7. With EDS and Kossel technique determined crystal phases Au_4Ti, Au_2Ti, and AuTi3 with plotted crystal orientation shown within the diffusion zone of the corresponding scanning backscattered electron micrograph (specimen 3) [48].

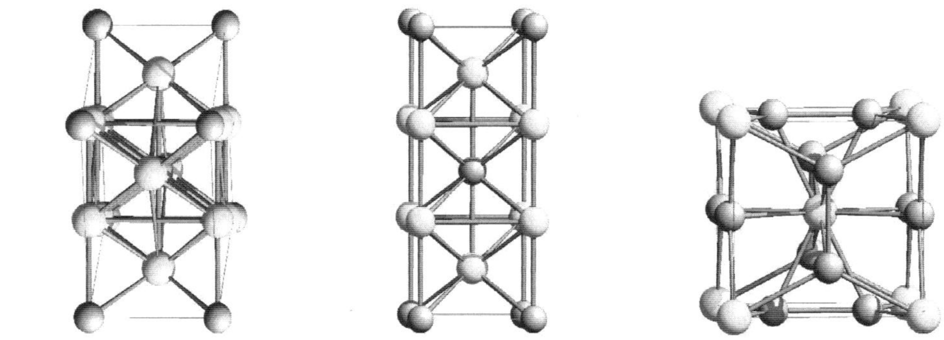

Figure 8. Spherical model representation of the relation between the most "open" crystal orientations of grains and the growth or diffusion direction of the crystal phases, which were found [48] (Au – yellow; Ti – grey). A similar dependence was proved in the Cu-Sn diffusion system also in the group of Däbritz *et al.* [50].

Table 6. Overview of the crystallographic data of the crystal phases.

Designation	Crystallographic data of the Au-Ti crystal phases [48]
Au_4Ti	tetragonal body centred, Ni_4Mo type, space group: I4/m (No. 87)
Au_2Ti	tetragonal body centred, $MoSi_2$ type, space group: I4/mmm (No. 139)
$AuTi_3$	primitive cubic, A15, β-W or Cr_3Si type, space group: Pm $\bar{3}$ n (No. 223)

3. CCD detection of X-ray Kossel microdiffraction patterns

For all those interested in EBSD also the X-ray Kossel diffraction could be exciting as a further additional µ-diffraction method, which can be easily combined with EBSD. Since there exist no commercial system as far as the author do know, important detail information about the applied scintillator CCD area detector combination to observe X-ray Kossel diffraction patterns should be given in the following, which Däbritz and Langer started to develop in the year 1998 [15, 16, 18].

3.1. Basic information about the image detection, formats and the developed image software

Fundamental information about the experiences with CCD detection are summarised in the following:

- At least a resolution of 1,280 ×1,024 pixels ($\approx$ 1.3 megapixel) of the CCD detector is required;
- 8-bit images are insufficient for Kossel patterns, at least 12-bit (4,096 grey values) or higher are needed;
- A 16-bit bitmap format (*.b16) is used to save the 12-bit bitmap CCD Kossel patterns, which is built up like the standard 8-bit image format, whereby the additional 4 bits are used for a text documentation for all observation parameters and a description of the sample;
- A special software is required in order to observe and process the digital images, which was developed in-house in the programming language C++ and is called "SENSICAM", whereby only the programming of the observation of the Kossel patterns was on the basis of a developer kit belonging to the CCD detector;

- The CCD area detector combined with the developed software allow an integration time of at least 1 - 5 min for X-ray Kossel patterns and more for special high-quality patterns (typical 6 – 8 min);
- Similar to EBSD the software controls the SEM CamScan CS44 to switch between scanning for the Kossel background pattern and the concentration of the electron beam on one spot to generate the Kossel diffraction pattern;
- In order to see the Kossel lines, it is necessary to subtract or to normalise the 12-bit bitmap Kossel patterns from the Kossel background pattern by analogy to the procedure for EBSD, see Figure 9;
- Identical with EBSD for polycrystalline samples, a scanning over different orientated grains is used for the background pattern, whereas for monocrystals the sample is rotated during the integration time of the background pattern to blur the Kossel lines, as can be seen on a Fe monocrystal in Figure 9;
- It is particularly important to achieve a perfect agreement between Kossel and background pattern by maintaining especially the same absorption current on average;
- For an excellent Kossel pattern, the integration time must be optimised by the grey value histogram, which should be as wide as possible, but the development of the Gaussian shaped peak on the right side has to be avoided, since it is a sign of a too highly exposed pattern, see Figure 10, which is caused by scattered electrons that appear at long integration times in dependence of the absorber foil thickness for the backscattered electrons.

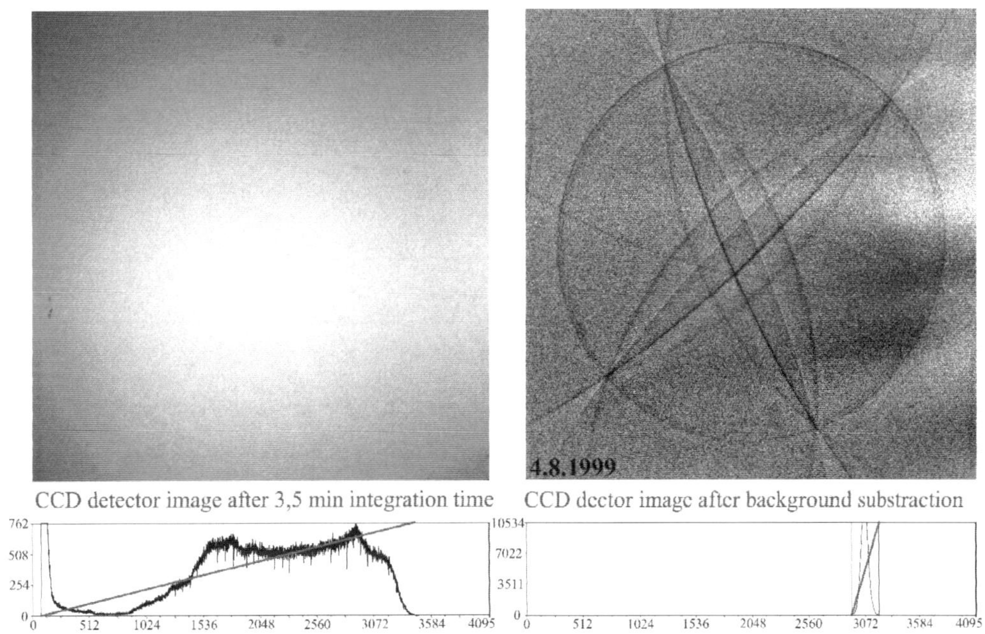

Figure 9. Comparison between Kossel line CCD pattern left and background subtracted Kossel pattern on the right side of a iron monocrystal at the [110] pole with corresponding grey value histogram in each case (acceleration voltage: 25 kV, absorption current: 0.1 µA, observed with phosphor type P43 and 3 gold foils each with 200 nm thickness were used between sample and screen in order to avoid the backscattered and diffracted electrons).

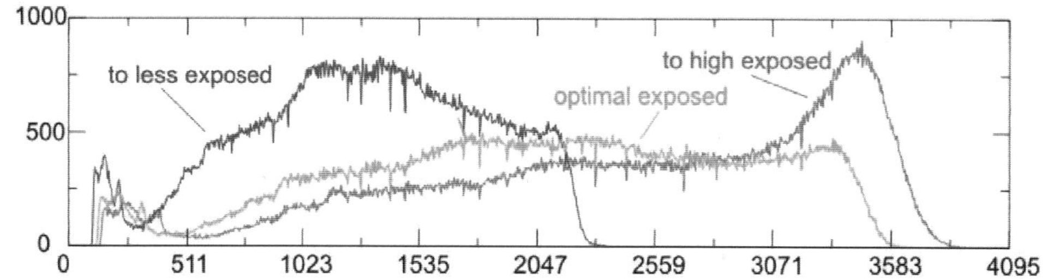

Figure 10. Grey value histogram of a too less, too much and optimal exposed Kossel pattern.

3.2. Optimisation of the phosphor scintillator

Different phosphors were tested as listed in Table 7 and additional from the material P22G ZnS: Cu, Al, Au screens with varied superficial density produced. The optimal screen both for X-ray Kossel diffraction and EBSD was from the type P43 with the in Figure 11 presented transmission geometry with Al reflex layer and the there given parameters.

Table 7. Arrangement of tested phosphors (www.apace-science.com/ast/phsophor.htm).

Type	Composition	Maximum emission λ/nm	Decay time to 10 % intensity	Particle size/µm	Energetical efficiency/%
P20	ZnCdS: Ag, Cu	530 (green)	80 µs	2.5 – 10	22
P22G	ZnS: Cu, Al, Au	540 (green)	35 µs	4 – 10	22
P43	Gd_2O_2S: Tb	545 (green)	1.5 ms	2.5 – 25	15

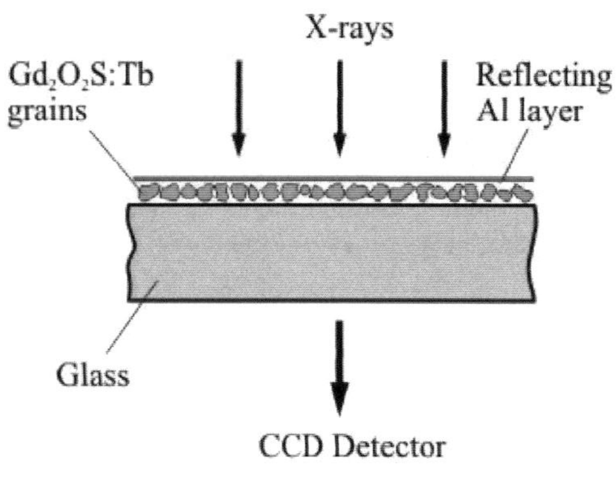

Type: Phosphor scintillator P43: Gd_2O_2S:Tb	
Superficial density	20 mg/cm^2
Average particle size	4 µm
Al reflection layer	50 nm
Glass substrate size	(100× 80× 2) mm^2
Usable phospor area	(98× 78) mm^2
Effective used section	(90× 70) mm^2

Figure 11. Schematic setup and data for X-ray Kossel diffraction optimised screen.

3.3. Filter foils for the X-ray Kossel pattern observation
For the Kossel pattern detection, between one (for a fast detection of 60 s) and up to three gold filter foils (high quality) were used between sample and screen; each with a thickness of 200 nm in order to avoid backscattered and diffracted electrons to form EBSP. Also a 30 μm aluminium foil was used for single excellent Kossel pattern, but this leads to a bit longer integration times.

4. Extension of the Kossel investigations to a further material class of ferromagnetic shape memory alloys

4.1. Introduction to the material of the system Co-Ni-Al
To the best of my knowledge, only one conventional thermal shape memory alloy of the system Cu-Al-Be has been investigated using the Kossel method to date and is only documented in a few publications [51, 52]. In 1996, the magnetostriction was measured by Vail'ev *et al.* [53] in a Ni_2MnGa single crystal and he suggested the practicability of controlling the size of crystals by an external magnetic field. The demonstration of such a so-called ferromagnetic shape memory alloy (FSMA) was performed by Ullakko *et al.* [54]. This involves Ni_2MnGa, a Heusler-type alloy. Nowadays FSMA's show clear changes with variation of an external magnetic field. Thereby changes in length of more than 10 % are possible. The Co-Ni-Al system attracts special attention as a potential candidate for the FSM effect due to certain features such as good oxidation resistance, low density and appreciable ductility at room temperature. The crystals were grown by the Bridgman method in [100] direction at different pulling rates [55, 56]. The present work studied austenitic [57] single crystals with nominal chemical composition $Co_{38}Ni_{33}Al_{29}$ by the X-ray Kossel diffraction and EBSD. Here only a short overview is given in the context of 100 years of the prediction of Kossel interferences. For the main Kossel experiments and the discussion more in detail please see in the parallel published paper of Langer, Potapov, and Kopeček [57].

4.2. Scanning electron microscopy and energy dispersive X-ray spectroscopy measurements
Figure 12 shows a backscattered electron micrograph of a typical sample with the designation AUST1 [57]. In this investigated high-temperature austenitic state, we observe a two-phase structure of a matrix and second crystal phase of precipitates, which are marked with these designations in black and white, respectively. In order to investigate whether the cause of the lower backscatter coefficient η at the grain boundaries in the matrix range can be explained with a change of the local chemical composition, a lot of energy-dispersive X-ray spectrometry (EDS) mappings of Co-Kα, Ni-Kα and Al-Kα were carried out, for one example see Figure 12. The EDS maps shows no clear change of composition close to the phase boundary in the matrix. Therefore, it only may be assumed, that the areas appear darker due to a better channelling effect on the basis of a higher crystal grade at the grain boundaries in the matrix range.

4.3. Investigation of the crystal structure by means of Kossel X-ray diffraction
In the ternary alloy Co-Ni-Al both the Co-K$\alpha_{1/2}$- and the Ni-K$\alpha_{1/2}$-radiation contributes to the Kossel diffraction, since the Al-Kα_1-radiation is clearly to long wave to fulfil the Bragg equation. Therefore, all Kossel lines appear, in this case, twice which means there are two identical reflection systems parallel to each other, see Figure 13, showing a typical indexed Kossel pattern in the matrix range, compare also with a further Kossel diffraction pattern of the main crystal phase (matrix) in [57]. The observation parameters of this pattern were the following: sample 60° tilted, accelerating voltage of 20 kV, the absorption current was with 1.45 μA relative high in this case and, therefore, the exposure time is only 180 s and only one 30 μm thick aluminium foil was used to avoid the backscattered and diffracted electrons. Although the intensity is separated for electron-excited Kossel X-ray interferences by 38 % for the Co-K$\alpha_{1/2}$-radiation as well as 33 % for the Ni-K$\alpha_{1/2}$-radiation, whereby the 29 at% Al atoms with the Al-K$\alpha_{1/2}$-radiation are not able to contribute, as discussed above, the Kossel pattern on the alloy

Co-Ni-Al surprisingly have such an excellent quality. Both the reflection systems are marked only for the {1 1 2} reflection type, whereby the Ni-radiation leads for each reflection to a larger Kossel cone opening angle than the Co-radiation. Only one of two double reflections is indexed in each case. As a result of the evaluation of a variety of Kossel patterns, the crystal structure in the matrix range could be determined with deviations independently to previous measurements with other methods, where an ordered matrix with the CsCl-structure type B2, space group $Pm\bar{3}m$ (No. 221), was found in the literature [58], which can be assigned as crystal phase β [57], see sphere model of Figure 13 on the right. The red marked reflections in the simulation of Figure 13 can be seen better.

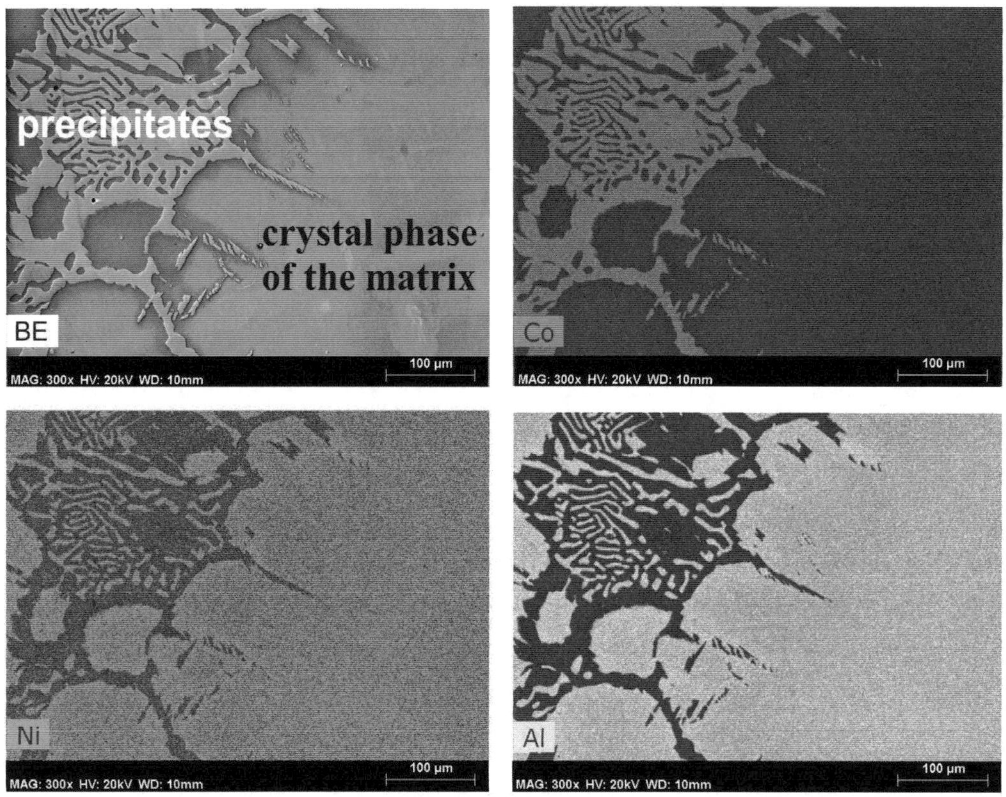

Figure 12. SEM micrograph of an austenitic $Co_{38}Ni_{33}Al_{29}$ alloy using backscattered electrons with both marked main crystal phase of the matrix and precipitates with corresponding differently coloured EDS maps of the qualitative content of Co, Ni and Al show no composition change close to the phase boundary in the matrix.

The following reflection types were observed in Kossel patterns of the matrix: {011}, {002}, {112}, {022}, and {103}. Thereby, five additional superstructure Kossel lines of the type {001}, {111}, {012}, {122} and {003} are needed to prove the simple cubic structure for the structure type B2. Not all superstructure Kossel lines could be clearly observed [57]. Only the following lines could be unambiguous documented: types {001} and {012}, but the types {111}, {122} and {003} could not be detected [57]. A possible reason for this is that the ternary B2 phase $(A,C)_{1-x}B_x$ has with $(Co,Ni)_{71}Al_{29}$

IOP Conf. Series: Materials Science and Engineering **1324** (2025) 012008 doi:10.1088/1757-899X/1324/1/012008

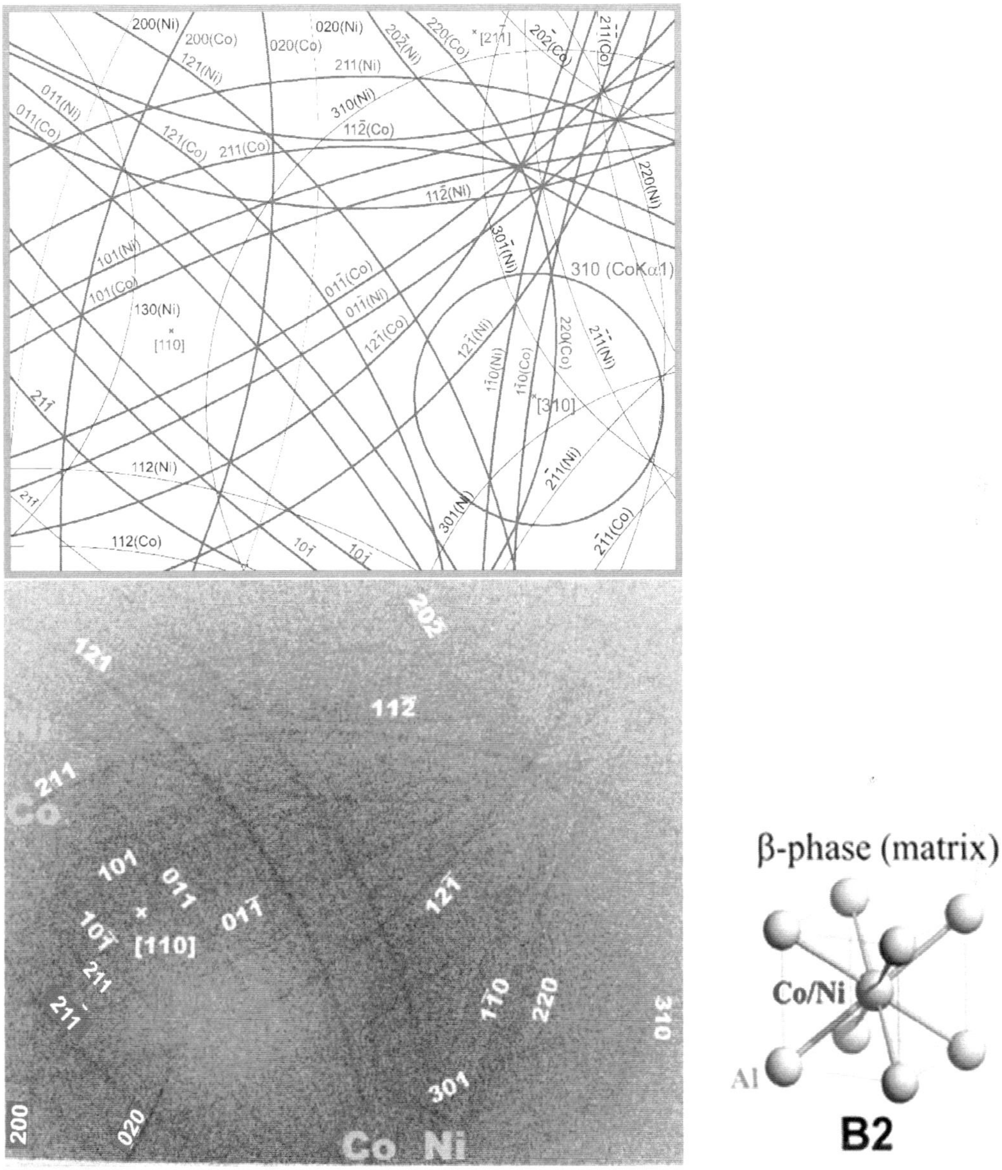

Figure 13. Left: Indexed Kossel pattern from the matrix of sample AUST1 with the corresponding simulation in gnomonic projection above [57], whereby the circular (310) Kossel reflection is remarkable, weakly seen at the bottom right, and is due to the small opening angle very sensitive to lattice constant changes and could therefore be useful for precise local lattice constant determinations in the micro-range. Right: Sphere model of the matrix β-phase with expected B2 structure: only the elements Co, Ni coloured in blue are statistically distributed.

not the ideally stoichiometric composition, then again the contrast ratios were relatively difficult in the corresponding Kossel patterns. The deviations of the ideal stoichiometric composition are very strong and amount to approximately 20 %. The perpetuation of the B2-structure is only possible by the formation of an enormous number of so-called constitutional crystal defects, whereby two kinds of point defects are important for ordered alloys: antistructure atoms and vacancies [59, 60], which can possibly explain the missing reflections.

5. Conclusions and outlook

With the applications and results, which were shown and discussed, the great performance of the Kossel X-ray μ-diffraction was demonstrated. Especially first Kossel measurements on a potential candidate for the ferromagnetic shape memory effect were presented. Moreover, in the last 20 years a few progress was achieved to overcome the following major disadvantages: Not only the CCD area detector replaced the X-ray film, but also the first successful steps have been taken in direction of the automated indexing of X-ray Kossel patterns by Maurice and Fortunier (2008) in Saint-Etienne, France [61]. Besides, these, a further new approach lies in the application of focal curves by Henschel and Bauch (2013) [62] in Dresden, Germany. Moreover, new Kossel simulation programmes were developed by different groups, especially KSLSTRAIN by Morawiec [63] in Krakow, Poland, in 2016, and by Bortel *et al.* in Budapest, Hungary [64] in 2021, which notably simplified and extended the evaluation.

Acknowledgments

I would like to thank the colleagues Carsten Schurig for his support with the development of the enhanced SensiCam-Control software 20 years ago, Leonid P. Potapov, Katerina Krátká, M.N. Melo for their experimental help at measurements, Kossel simulations and indexing on Co-Ni-Al, and especially Jaromír Kopeček for the crystal growing, the introduction to the field of FSMA's and for his contributions to the results presented here regarding the Co-Ni-Al alloy.

References

[1] Kossel W 1924 *Z. Physik* **23** 278
[2] Kossel W, Loeck V and Voges H 1935 *Z. Physik* **94** 139
[3] Kossel W 1937 *Erg. exakt. Naturwiss.* **16** 295-352
[4] Borrmann G 1935 *Naturwiss.* **23** 591; 1936 *Ann. Phys. (Leipzig)* **27** 669
[5] Geist V and Flagmeyer R 1974 *phys. stat. sol. (a)* **26** K1
[6] Roberto J B, Batterman B W, Kostroun V O, Appleton B R 1975 *J. Appl. Phys.* **46** 936-937
[7] Ullrich H-J, *et al.* 1994 *Nucl. Instrum. Methods Phys. Res. A* **349** 269
[8] Schetelich C, Weber S, Geist V, Schlaubitz M, Ullrich H-J, Kek S and Krane H G 1995 *Nucl. Instrum. Methods. Phys. Res. B* **103** 236
[9] Langer E, Däbritz S, Hauffe W and Haschke M 2005 *Appl. Surf. Sci.* **252** 240
[10] Langer E, Haschke M and Däbritz S 2008 *Mikrochim. Acta* **161** 455
[11] Peyrusse O, Jonnard P, Le Guen K and André J-M 2020 *Phys. Rev. A* **101** 013818
[12] von Laue M 1960 Röntgenstrahl-Interferenzen. [Frankfurt am Main, Germany: Akademische Verlagsgesellschaft]
[13] Goehner M and Michael J R 1995 *Adv. X-ray Anal.* **38** 539
[14] Ullrich H-J, Brechbühl J, Bauch J, Lin H, Bartunik H-D, Bösecke P and Bourenkov G 1998 *HASYLAB annual report.* [Hamburg, Germany: HASYLAB] 887-888
[15] Däbritz S, Langer E and Hauffe W 1999 *J. Anal. At. Spectrom.* **14** 409
[16] Däbritz S, Langer E and Hauffe W 2001 *Appl. Surf. Sci.* **179** 38
[17] Langer E and Däbritz S 2010 *IOP Conf. Series: Mat. Sci. Eng.* **7** 012015-1-15
[18] Langer E, Däbritz S, Schurig C and Hauffe W 2001 *Appl. Surf. Sci.* **179** 45
[19] Wilkinson A J, Meaden G and Dingley D J 2006 *Ultramicroscopy* **106** 307
[20] Wilkinson A J, Meaden G and Dingley D J 2006 *Mater. Sci. Technol.* **22** 1271

[21] Winkelmann A, Schröter B and Richter W 2004 *Phys. Rev. B* **69** 245417
[22] Winkelmann A, Trager-Cowan C, Sweeney F, Day A P and Parbrook P 2007 *Ultramicroscopy* **107** 414
[23] Nolze G, Grosse C and Winkelmann A 2015 *J. Appl. Crystallogr.* **48** 1405
[24] Ullrich H-J, Thiele K, Däbritz S, Schreiber H and Götze K 1972 *Krist. Techn.* **7** 1153
[25] Nolze G, Winkelmann A, Cios G and Tokarski T 2021 *Mater. Charact.* **175** 111040
[26] Bauch J, *et al.* 1999 *Cryst. Res. Technol.* **34** 71
[27] Brechbühl J, Bauch J and Ullrich H-J 1999 *Cryst. Res. Technol.* **34** 59
[28] Böhling M, Bauch J and Ullrich H-J 2009 *Cryst. Res. Technol.* **44** 1185
[29] von Laue M 1935 *Naturwiss.* **23** 373
[30] von Laue M 1935 *Ann. Phys.* **23** 705
[31] Ott H 1938 *Ann. Phys.* **31** 264
[32] Blau W, Schreiber H, Schulze G E R and Ullrich H-J 1967 *Naturwiss.* **54** 535
[33] Stephan D, Blau W, Ullrich H-J and Schulze G E R 1974 *Krist. Techn.* **9** 707
[34] Stephan D, Ullrich H-J and Schulze G E R 1976 *Krist. Techn.* **11** 475
[35] Brümmer O and Nieber J 1975 *Mikrochimica Acta* **Suppl. 6** 345
[36] Hutton J T, Trammell G T and Hannon J P 1985 *Phys. Rev. B* **31** 743
[37] Faigel G, Bortel B and Tegze M 2016 *Sci. Rep.* **6** 22904
[38] Brümmer O, Beier W and Nieber J 1973 *phys. stat. sol. (a)* **20** K119
[39] Geist V and Flagmeyer R 1974 *phys. stat. sol. (a)* **26** K1
[40] Ullrich H-J, Schatt W, Däbritz S and Geist V 1977 *Microchim. Acta* **Suppl. II** 167
[41] Nolze G, Geis V, Wagner G, Paufler P and Jurkschat K 1990 *Z. Kristallogr.* **193** 111
[42] Ullrich H-J, Herenz A, Friedrich E, Schatt W, and Döring C 1983 Mikrochim. Acta I 175
[43] Herenz A, Ullrich H-J and Däbritz S 1984 in: *Proc. 6 th Microprobe Conf.* [Dresden, Germany: Physical Society of the GDR] 203
[44] Däbritz S 1974 *PhD thesis.* [Dresden, Germany: Dresden University of Technology]
[45] Ullrich H-J 1967 *PhD thesis.* [Dresden, Germany: Dresden University of Technology]
[46] Ullrich H-J, Däbritz S, Schreiber H 1968 in: *Proc. 5th Int. Congr. X-ray Optics and Microanalysis.* (Möllenstedt G and Gaukler K H; Eds.) [Tübingen, Germany] 406-411
[47] Richter K, Keiter K, Bergner D, Müller A, Raub C J and Ott D 1997 *Mikrochim. Acta* **125** 115
[48] Kurt R, Däbritz S, Hauffe W, Bergner D and Richter K 1997 *Defect and Diffusion Forum* **143-147** 609
[49] Langer E, Kurt R and Däbritz S 1999 *Cryst. Res. Technol.* **34** 801
[50] Däbritz S, Hauffe W and Kurt R 1997 *Mikrochim. Acta* **125** 3-12
[51] Bouscaud D, Berveiller S, Pesci R, Patoor E and Morawiec A 2014 *Adv. Mater. Res.* **996** 45
[52] Pesci P, Inal K, Berveiller S, Patoor E, Lecomte J and Eberhardt A 2006 *Mater. Sci. Forum* **524-525** 109-114
[53] Vasil'ev A N, Klestov S A, Levitin R Z and Snegirev V V 1996 *J. Exper. Theor. Phys.* **82** 524
[54] Ullakko K, Huang J K, Kantner C, O'Handley R C and Kokorin V V 1996 *Appl. Phys. Lett.* **69** 1966
[55] Kopecek J, Sedláková-Ignácová S, Jurek K, Jarošsová M, Drahokoupil J, Šittner P and Novák V 2009 in: *ESOMAT 2009 - 8th Eur. Symp. Martensitic Transformations.* 02013
[56] Kopeček J, Jurek K, Jarošová M, Drahokoupil J, Sedláková-Ignácová S, Šittner P and Novák V 2010 *IOP Conf. Series: Mat. Sci. Eng.* **7**, 012013
[57] Langer E, Potapov L and Kopeček J 2024 *IOP Conf. Series: Mat. Sci. Eng.*, this volume
[58] Hubert-Protopescu M and Hubert H 1991 *Aluminium-cobalt-nickel ternary alloys: a comprehensive compendium of evaluated constitutional data and phase diagram. Vol. 4: Al-Cd-Ce to Al-Cu-Ru.* (Petzow G and Effenberg G, Eds.) [Weinheim, Germany: VCH] 234
[59] Grün A, Henig E-T and Sommer F 1998 *Z. Metallk.* **89** 591
[60] Breuer J, Sommer F and Mittemeijer E J 2001 *Metall. Mater. Trans. A* **32** 2157

[61] Maurice C and Fortunier R 2008 *J. Microscopy* **230** 520
[62] Henschel F and Bauch J 2013 *J. Geom. and Graphics* **17** 205
[63] Morawiec A 2016 *J. Appl. Cryst.* **49** 322
[64] Bortel G, Tegze M and Faigel G 2021 *J. Appl Cryst.* **54** 123

Investigations of the ferromagnetic shape memory alloy Co-Ni-Al by X-ray Kossel microdiffraction

E Langer[1,2], L P Potapov[2] and J Kopeček[3]

[1] Dresden University of Technology, Institute for Semiconductors and Microsystems, 01062 Dresden, Germany
[2] Dresden University of Technology, Institute of Solid State and Materials Physics, 01062 Dresden, Germany
[3] Academy of Sciences of the Czech Republic, Institute of Physics, Na Solvance 2, 18221 Prague, Czech Republic

E-mail: langer@physik.tu-dresden.de

Abstract. The ternary compound Co-Ni-Al was investigated by the X-ray Kossel microdiffraction as a member of this material class of ferromagnetic shape memory alloys for the first time. Thereby monocrystals with the nominal chemical composition $Co_{38}Ni_{33}Al_{29}$ were initially studied in the austenitic phase, in order to contribute to a better understanding of the complex behaviour of this material by means of a further diffraction method with a high informational content. While in the present two phase structure the precipitates could be clearly identified and confirmed as face-centred cubic cobalt solid solution by the Kossel diffraction, whereas for the expected B2-structure only 2 of the needed 5 superstructure reflections could be proved by means of the Kossel technique. A possible explanation lies besides the reason of a weak reflection intensity in a very strong discrepancy of 21 % of the ideal stoichiometric composition of $(Co, Ni)_{50}Al_{50}$ in the investigated $(Co, Ni)_{71}Al_{29}$. As known from the literature to maintain the B2-structure under such conditions is only possible by the formation of an enormous number of so-called constitutional crystal defects, which could be used to explain the non-verified superstructure reflections. Moreover, the local lattice constant of the main crystal phase β was determined to a = (0.2855 ± 0,0002) nm by the Kossel technique.

1. Introduction

First, we must distinguish between the conventional thermal shape memory effect and the ferromagnetic shape memory effect, where the magnetic field plays the decisive role. Only one conventional shape memory alloy (SMA) of the system Cu-Al-Be has been investigated using the Kossel method [1, 2] to date, and was only published in a few publications [3, 4]. Not only thin SMA, but also ferromagnetic SMA (FSMA) films have attracted attention for potential applications as smart actuators in microelectromechanical systems (MEMS), for example as alternatives for electroactive polymer (EAP) actuators [5].

The ferromagnetic shape memory effect in Ni-Mn-Ga alloys has been the subject of much attention, mainly as the prolongation of the single crystal grew from original 0.2 % in 1996 up to 10 % in the year 2000 [6, 7]. The Co-Ni-Al system appeared to be another example of this family, especially since it has certain advantageous properties such as appreciable ductility at room temperature, a relative low density, and good oxidation resistance. Nevertheless, the magnetic behaviour of cobalt alloys is different than in the case of Ni-Mn-Ga and a significant effect of the magnetic forces on microstructures was not observed. Moreover, the Co-Ni-Al alloys have a two-phase structure even at the austenitic, high-temperature state. It is expected that such a structure consists of a B2-matrix and A1-particles, whereby the latter means a face-centred cubic (fcc) cobalt solid solution. The martensitic transformation undergoes just B2 ordered matrix, whereas the A1-particles remains untransformed [8]. The complicated microstructure in annealed and quenched samples led to an instability of the cubic lattice and a significant sensibility to stress induced transformation. Such behaviour was proven recently [9], which explain the formed observations pointing that the equilibrium martensitic shape memory take place at around 200 K in the $Co_{38}Ni_{33}Al_{29}$ alloy [10].

The present work studied austenitic single crystals with nominal chemical composition $Co_{38}Ni_{33}Al_{29}$ by the X-ray Kossel diffraction. The principle of the Kossel technique is sketched in Figure 1 for this ternary system Co-Ni-Al. In this case the Bragg equation $2d_{hkl}sin\Theta_{hkl} = \lambda_{specimen}$ has to be applied to the wavelengths λ's of Co-, Ni-, and Al-Kα radiation of the excited elements cobalt, nickel and aluminium of the investigated specimen for the Kossel diffraction. This means for the cubic crystal system with the lattice plane distances $d_{hkl} = \frac{a}{\sqrt{h^2+k^2+l^2}}$ and α as lattice constant, $sin\Theta_{hkl} = \frac{\lambda_{specimen}}{2a}\sqrt{h^2+k^2+l^2} \leq 1$ must be valid, because of the trigonometric function $sin\Theta_{hkl}$. Both for $\lambda_{specimen} = \lambda_{Co-K\alpha_1} = 0.1788996\,nm$, and $\lambda_{specimen} = \lambda_{Ni-K\alpha_1} = 0.1657930\,nm$ [11] and with the lattice constant α = 0.2861 nm for CoAl (B2) [12] $sin\Theta_{103} = 0.9887 \leq 1$ and $sin\Theta_{103} = 0.9163 \leq 1$, respectively, is fulfilled even for this higher order reflection in either case. The Al-$K\alpha_1$ radiation with a wavelength of $\lambda_{specimen} = 0.8339514\,nm$ is clearly more long wave and already the calculation of a reflection of the lowest possible order for the simple cubic case gives $sin\Theta_{001} = 1.46 \geq 1$, which means, that the Bragg relation cannot be fulfilled for Al-radiation. Therefore, the Kossel reflections appear in characteristic doublets of inner Co and outer Ni Kossel lines.

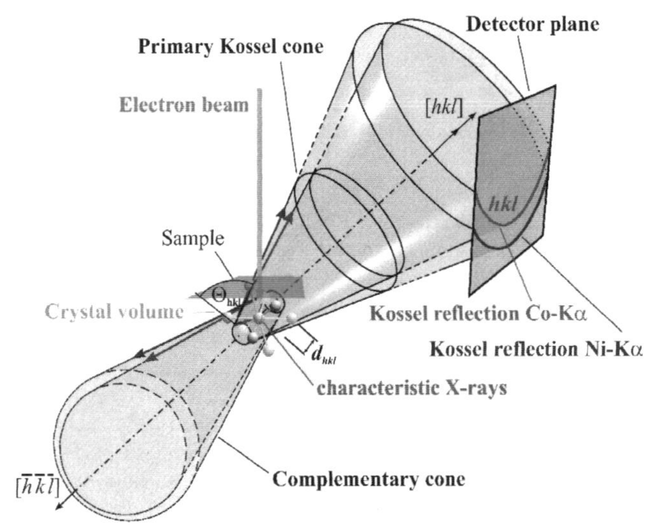

Figure 1. Principle of the Kossel technique sketched on the example of Co-Ni-Al.

2. Experimental

2.1. Sample preparation

The polycrystalline alloy $Co_{38}Ni_{33}Al_{29}$ was used to prepare single-crystals. The crystals were grown by the Bridgman and floating-zone methods in [100] direction [13, 14]. The low pulling rates (below 10 mm·h^{-1}) led to the disintegration of the crystal body into an two-phase mantle and a single-phase core, which was an undesirable structure [14]. The higher pulling rates in Bridgman method gave approximately the same composition of the main crystal phase and of particles than the second crystal phase, although the structure along the crystal was not homogeneous [13]. Finally, the pulling rate 10^4 mm·h^{-1} produced a homogeneous distribution of particles of the second crystal phase within the main crystal phase. Such a microstructure is the only one suitable for the repeatable sample preparation. Table 1 summarises the preparation conditions and heat treatment for the investigated specimens. The samples were annealed and quenched in ice water immediately after the thermal treatment, for details see in Table 1. Conventional metallographic polishing using colloidal silica at the final step was utilised for the sample preparation.

Table 1. Overview of the prepared and investigated Co-Ni-Al samples.

Sample Designation	Alloy	Preparation conditions	Pulling rate	Heat treatment
AUST1	$Co_{38}Ni_{33}Al_{29}$	1530 °C	104 mm/h	1275 °C, 4 h
BM3	$Co_{38}Ni_{33}Al_{29}$	1530 °C	37.4 mm/h	1275 °C, 4 h

2.2. Instrumentation

The microstructure was characterised using a scanning electron microscope (SEM) of the type JEOL JSM-6610LV equipped with LaB_6 cathode. The energy-dispersive X-ray spectrometry (EDS) was carried out with a 10 mm^2 silicon drift detector (SDD) Bruker X-Flash 6|10 within the same SEM. This SDD generation provides a maximum energy resolution of 121 eV at Mn-Kα and allows count rates of over 1 million impulses per second for fast linescans and mappings. All Kossel X-ray microdiffraction experiments were performed at a SEM CamScan CS 44 also with a LaB_6 cathode and the possibility of a maximum acceleration voltage of 40 kV. A combination of a scintillator and a SensiCam CCD (charged coupled device) area detector with an objective lens system was applied for the observation of Kossel patterns, which was developed by Däbritz and Langer [15, 16, 17]. The PCO SensiCam 370XL CCD detector has the following specifications: digital 12 bit, 2/3 inch sensor (Sony ICX085AL, 1300×1030 pixel), two-stage Peltier cooled to -12 °C with scientific grade for the detection of extremely faint signals, usable resolution of 1280×1024 pixel, a pixel size on the chip of 6.7 µm×6.7 µm, absolute no pixel errors, a quantum efficiency of typical 40 % (monochrome, $\lambda = 500$ nm), fibre optic link via fibre-glass-cable to decrease electronic noise, long exposure times from 1 ms – 1,000 s. The high resolution X-ray phosphorus scintillator P43: Gd_2O_2S:Tb with a maximum of the light emission at $\lambda = 545$ nm (green), an average particle size of 4 µm and a superficial density of 20 mg/cm^2 on a glass substrate with an aluminium reflecting layer of a thickness of 50 nm was used. The pixel size on the scintillator was 70 µm and the shortest distance between electron focus and screen (pattern centre) was approx. 75 mm. Usually up to three separated 200 nm thin gold foils or sometimes one 30 µm thick aluminium foil was used to avoid the backscattered and diffracted electrons. The self-developed simulation programme KOPSKO2.1 [18] was used to evaluate the diffraction patterns.

3. Results and discussion

3.1. Characterisation of the microstructure by scanning electron microscopy

Since the samples were prepared at different pulling rates, besides the standard observation using light microscopy, the individual characterisation of the microstructure of each sample by means of SEM was necessary. Figure 2 shows a typical backscattered electron micrograph of the sample AUST1. Most of the crystal volume is occupied by a homogeneous crystal phase, which is, therefore, referred to as "matrix" in this paper. Another structural component seems to originate from interdendritic melt remains. Its morphology resembles an eutectic mixture of two different solid phases with a labyrinth-like structure. In addition, micro-scale precipitates [14] are observed in some of the samples, embedded within the matrix. All precipitates of the expected second phase share a similar backscatter coefficient η, which is higher than that of the matrix, which makes the second phase to appear brighter in all SEM images. Only one exemplary SEM micrograph of sample AUST1 is shown, since the electron microscopic investigations of the second sample BM3 can be described as very identical to the above discussed properties. Only for the sample AUST1 further characteristics were observed which can be partially classified more on the macroscopic scale of several mm. Although the sample AUST1 was grown as single crystal, the variance of the orientation contrast of backscattered electrons suggests slightly different crystallographic orientation in different areas of the matrix, probably due to their dendritic growth. As a result, the microstructure looks much like mosaic crystal blocks separated by either large lamella-like eutectic regions (see Figure 2) or thin borders of ≤ 10 µm width, which consist of the second crystal phase.

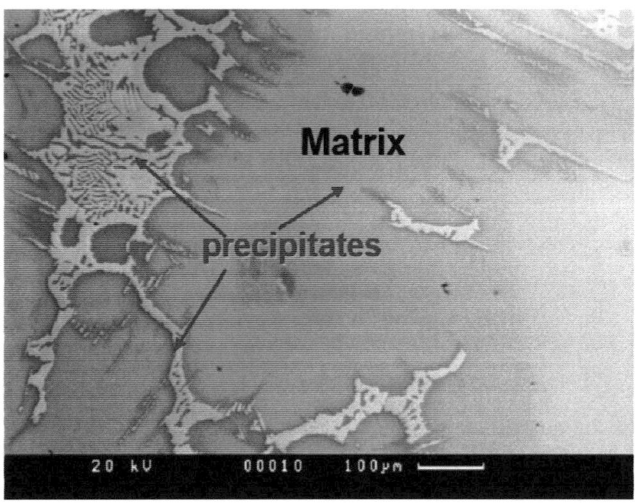

Figure 2. The SEM micrograph of the austenitic $Co_{38}Ni_{33}Al_{29}$ alloy of sample AUST1 using backscattered electrons shows a two-phase microstructure: a largely homogeneous occupied range in a dark grey material contrast, which is referred to as matrix and a second crystal phase of precipitates with bright contrast, that means a higher backscatter coefficient η, which are marked with blue arrows.

On the whole it can be stated that in this investigated high-temperature austenitic state a two-phase structure of a matrix and a second crystal phase can be observed in both samples.

3.2. Crystal phase analysis by energy-dispersive X-ray spectrometry

In order to confirm the two phase structure several EDS mappings of the characteristic X-rays of the elements Co-Kα, Ni-Kα and Al-Kα were carried out on both samples AUST1 and BM3, with subsequent phase analysis, see Figure 3. The resulting phase analysis visualizes a chemically distinct matrix phase (shown in red) and the second phase (coloured in yellow), as can be seen in Figure 3. This clearly demonstrates that this is a two-phase alloy. Moreover, it enables the identification of the interdendritic material as a mixture of both crystal phases (Figure 3a) and the matrix-embedded particles as precipitates obviously from the same second crystal phase besides the matrix from the aspect of the chemical composition (Figure 3b).

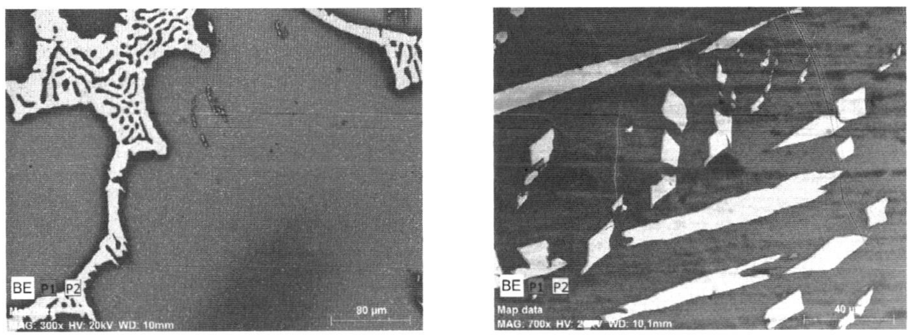

a) b)

Figure 3. EDS maps with phase analysis of two different morphologic structures within both samples. a) Selected region with matrix and eutectic mixture of sample AUST1. b) Example of precipitates in the matrix of sample BM3. The crystal phase of the matrix is displayed in red and the second crystal phase accordingly in yellow.

Since the EDS point measurements of the sample AUST1 would present not really new results compared with already detailed EDS measurements on samples with similar crystal growing parameter in the past by Kopeček *et al.* [13], we would focus more on the results of sample BM3, which shows especially a contamination, which was not investigated before. The corresponding precise EDS point measurements of sample BM3 in Table 2, with extremely high total counts of $5 \cdot 10^6$ for point measurements, show that in comparison with the matrix the second phase is enriched with Co. It explains the higher backscatter coefficient within the second phase. The nickel content is almost comparable in both phases. Also, low contamination amounts of iron with statistical uncertainties of ≤ 1.0 at% were found only in sample BM3.

Table 2. Selected EDS results for the composition of matrix and the second crystal phase in the Co-Ni-Al sample BM3.

Position/Morphology	**Co** /At.%	**Ni** /At.%	**Al** /At.%	**Fe** /At.%
Matrix grid points 1–30	39.7(14)	33.9(10)	25.6(11)	0.8(5)
Matrix random points 1–10	39.5(15)	33.5(13)	25.8(16)	1.2(7)
Phase points 1–10	53.5(13)	30.2(11)	14.8(9)	1.5(9)
Mapping of the phase	38.5(37)	33.4(35)	26.9(57)	1.2(53)
Mapping of the matrix	53.1(40)	29.3(34)	16.1(55)	1.6(51)

3.3. First Kossel diffraction studies of ferromagnetic shape memory alloys

3.3.1. Investigation of the matrix phase of Co-Ni-Al. As already discussed in detail in the Section 1, in such a ternary system Co-Ni-Al both the Co-K$\alpha_{1/2}$ of the cobalt atoms and the Ni-K$\alpha_{1/2}$ radiation of the nickel atoms contribute to the Kossel X-ray μ-diffraction, since the Al-Kα_1 radiation is clearly to long wave to fulfil the Bragg equation. Thus, the radiation of aluminium atoms does not contribute to the Kossel lines, but the elements Co, Ni and Al contribute together in the continuous unscattered background, reducing the contrast and making the detection of Kossel lines challenging. Since both the electron excited nickel and the cobalt atoms serve as X-ray lattice sources in the investigated sample, the Bragg equation is fulfilled equally for both wavelengths and, therefore, all Kossel lines appear in this case twice. Compared with conventional X-ray diffraction it would mean to use two external X-ray tubes Co and Ni with monochromatised Co-K$\alpha_{1/2}$ and Ni-K$\alpha_{1/2}$ radiation simultaneously. This means there are two identical reflection systems parallel to each other, please see Figure 4, showing a typical indexed Kossel pattern in the matrix range. Although the intensity is separated for electron excited Kossel X-ray interferences by 38 % for the Co-K$\alpha_{1/2}$ radiation as well as 33 % for the Ni-K$\alpha_{1/2}$ radiation, whereby the 29 at% Al atoms with the Al-K$\alpha_{1/2}$ radiation are not able to contribute, like discussed above, the Kossel pattern on the alloy Co-Ni-Al surprisingly have such an excellent quality. Both the reflection systems are marked only for the $10\bar{1}$ reflections, whereby the Ni-radiation leads for each reflection to a larger Kossel cone opening angle than the Co-radiation. Only one of the two double reflections is indexed in each case. As a result of the evaluation of a variety of Kossel patterns, the crystal structure in the matrix range could be determined independently to previous measurements on widely similar samples (but not the same specimen) with other methods, where an ordered matrix with the CsCl-structure type B2, space group $Pm\bar{3}m$ (No. 221), was found [19], which can also be assigned as crystal phase β, see crystal sphere model of Figure 5. Our results reveal that some deviations to a clear CsCl-structure type B2 could be observed and also not a simple body centred cubic structure can explain the measured reflection types, which should be discussed in detail next.

The following reflection types were observed in Kossel patterns of the matrix: {011}, {002}, {112}, {022}, and {103}, see Table 3. If these reflections would be present it could correspond to a body centred cubic structure, see the third column of Table 3. Thereby, five additional superstructure Kossel lines of the type {001}, {111}, {012}, {122} and {003} are necessary in order to prove the simple cubic structure for the structure type B2. Not all superstructure Kossel lines could be clearly observed. Only the following lines could be unambiguous documented: types {001} and {012}, but the types {111}, {122} and {003} could not be detected. A possible reason for this is that the ternary B2 phase $(A, C)_{1-x}B_x$ has with $(Co, Ni)_{71}Al_{29}$ not the ideally stoichiometric composition, then again the contrast ratios were relatively difficult in the corresponding Kossel patterns. The deviations of the ideal stoichiometric composition is with 21 % very strong. The perpetuation of the B2-structure is only possible by the formation of an enormous number of so-called constitutional crystal defects, whereby two kinds of point defects are important for ordered alloys: anti-sites and vacancies [20, 21], which can possibly explain the missing reflections. The circular (130) Co-Kα_1 Kossel reflection in Figure 4, weakly seen at the bottom, is due to its large Bragg angle extremely sensitive to lattice constant changes (which potentially may occur locally in the micro-range of alloys). Therefore, this sensitivity was very useful for just a graphical lattice constant determination by evaluating and comparing the distances of the circle to neighbouring reflections by slight variation of the lattice constant α in the corresponding simulation of Figure 4. In this way the local lattice constant of the main crystal phase β in sample AUST1 was known independently lattice constants determined to $\alpha = (0.2855 \pm 0{,}0002)$ nm with already a good precision by means of the X-ray Kossel technique, which were used then for all Kossel simulations of the matrix.

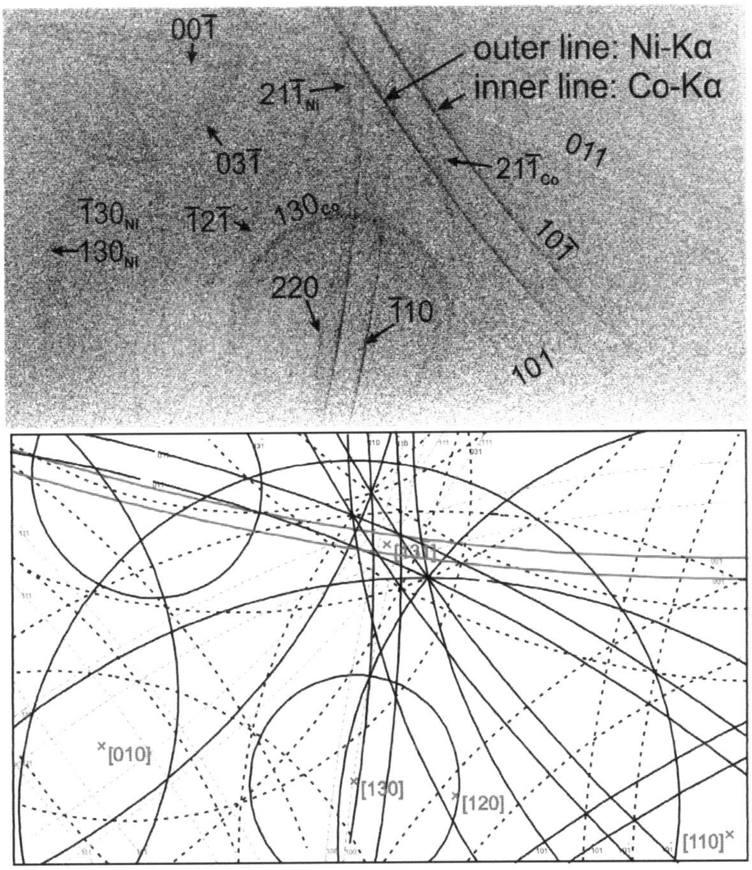

Figure 4. Exemplary indexed Kossel patterns with enhanced contrast, acquired within the matrix of sample AUST1, are displayed on top with corresponding simulations on the bottom. The typical acquisition parameters were: accelerating voltage of 25 kV, absorption current of 1.25 μA and exposure times of 300 s. Visible reflections are marked black, while reflections lacking enough contrast to be seen directly are represented with dotted lines. The $(00\bar{1})$ superstructure reflection of the {100}-type is marked magenta in the simulation. The circular (130) Kossel reflection shows a splitting in the two components of Co-Kα_1 (outside) and Co-Kα_2 (inside), which is remarkable.

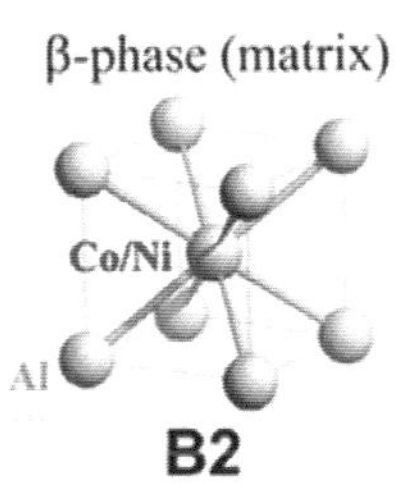

Figure 5. Sphere model of the β crystal phase of the matrix with B2-structure: Only Co and Ni atoms coloured in blue are statistically distributed.

Table 3. Overview of observed Kossel and superstructure reflections marked in blue.

$\sum h_i^2$	observed types of reflections	needed superstructure reflections for B2 type	reflections for bcc
1	{001}	{001}	
2	{011}	{011}	{011}
3		{111}	
4	{002}	{002}	{002}
5	{012}	{012}	
6	{112}	{112}	{112}
8	{022}	{022}	{022}
9		{122}	
9		{003}	
10	{103}	{103}	{103}

3.3.2. Kossel investigation of the second crystal phase. A variety of Kossel measurements in both samples AUST1 and BM3 of different points on the second crystal phase which appear brighter in the backscattered electron micrographs and corresponding to the EDS maps show very clearly only Kossel reflections of the face centeed cubic (fcc) structure type A1, space group $Fm\bar{3}m$ (space group No. 225). Therewith the crystal phase γ can also be assigned, as seen in Figure 6. These indicates the statistical distribution of the elements Co, Ni and Al within the lattice as shown in the crystal sphere model of Figure 7. Figure 6 shows only one example Kossel pattern of the sample AUST1. In this case three separated 200 nm thin gold foils were used to avoid the backscattered and diffracted electrons. Furthermore, no influence of the Fe impurities (see Table 2), which were only found in this sample BM3, was measured in the diffraction patterns of the β- and γ-phase in the sample BM3. Similar observed Kossel diffraction patterns at the sample BM3 could not be presented here, due to the limited space in this paper, whose evaluation provided the same results also for the second crystal phase

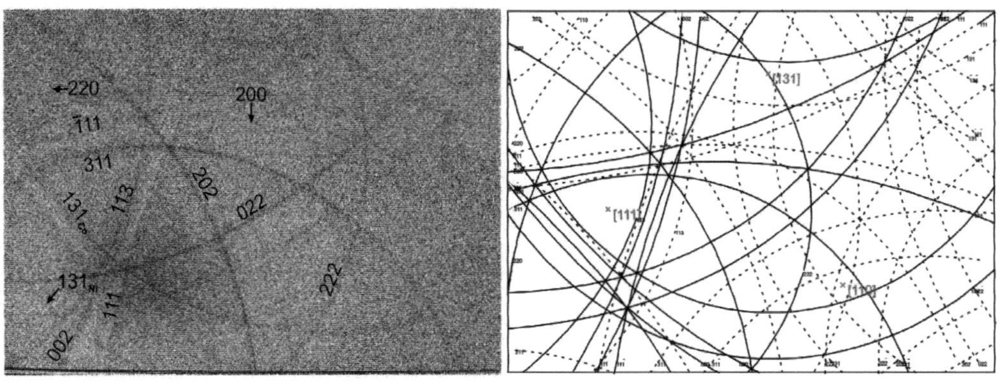

Figure 6. X-ray Kossel diffraction pattern of the Co-Ni-Al alloy in the second phase with corresponding Kossel simulation and indexing. The acquisition parameters were: an accelerating voltage of 20 kV, an absorption current of 1.25 µA, an exposure time of only 180 s and the sample was 60° tilted. Visible reflections are marked black, while reflections lacking enough contrast to be seen directly are represented with dotted lines.

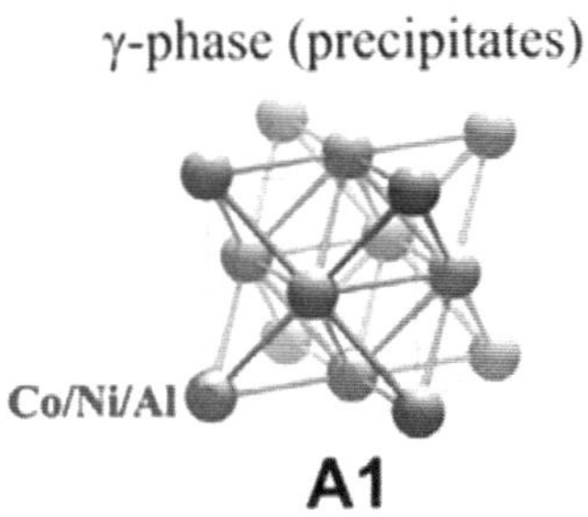

Figure 7. Sphere crystal model with statistical distribution of all elements Co, Ni and Al on the crystal lattice places of the A1-structure (fcc) of the measured γ-phase of the precipitates.

4. Conclusions

The microstructure of the alloy $Co_{38}Ni_{33}Al_{29}$ was investigated by SEM, EDS and first Kossel measurements on this potential candidate for the ferromagnetic shape memory effect were presented. On the whole, it can be stated that in this investigated high-temperature austenitic state of the Co-Ni-Al alloy a two-phase structure of a matrix and a second crystal phase can be proved in both samples with SEM, EDS and Kossel microdiffraction. The Kossel investigations of the crystal phase of the matrix could not explicitly prove a B2-structure, since three further necessary superstructure reflections could not be observed. The deviations of the ideal stoichiometric composition is approx. 20 %, very strong in the $(Co, Ni)_{71}Al_{29}$ alloy. The perpetuation of the B2-structure is only possible by the formation of an enormous number of so-called constitutional crystal defects, whereby two kinds of point defects are important for ordered alloys: anti-sites and vacancies [20, 21], which can possibly explain the missing or too weak superstructure reflections. Therefore, the Kossel measurements support the presence of a superstructure, which is more complex than the standard B2. The second crystal phase of the precipitates could be clearly confirmed as fcc, that means the A1-structure.

Acknowledgments

The authors would like to thank the colleagues Katerina Krátká for their experimental help. We acknowledge CzechNanoLab Research Infrastructure supported by MEYS CR (LM2023051).

References

[1] Kossel W, Loeck V and Voges H 1935 *Z. Physik* **94** 139-144
[2] Langer E and Däbritz S 2010 *IOP Conf. Ser.: Mater. Sci. Eng.* **7** 012015
[3] Bouscaud D, Berveiller S, Pesci R, Patoor E and Morawiec A 2014 *Adv. Mater. Res.* **996** 45
[4] Pesci P, Inal K, Berveiller S, Patoor E, Lecomte J S and Eberhardt A 2006 *Mater. Sci. Forum* **524-525** 109-114
[5] Franke M, Slowik I, Langer E, Leo K and Richter A 2020 *Surf. Interface Anal.* **52** 1163-1170
[6] Ullakko K, Huang J K, Kantner C, O'Handley R C and Kokorin V V 1996 *Appl. Phys. Lett.* **69** 1966-1968
[7] Heczko O, Sozinov A and Ullakko K 2000 *IEEE Trans. Mag.* **36** 3266-3268
[8] Kopeček J, Kopecký, Landa M and Heczko O 2013 *Mat. Sci. Forum* **738-739** 416-420
[9] Seiner H, Kopeček J, Sedlák P, Bodnárová L, Landa M, Sedmak P and Heczko O 2013 *Acta Materialia* **61** 5869-5876
[10] Heczko O, Seiner H, Sedlák P, Kopeček J and Landa M 2012 *J. Appl. Phys.* **111** 07A929
[11] Prince E (Ed.) 2004 *International tables for crystallography. 3rd edition. Vol. C: Mathematical, physical and chemical tables.* [London, UK: John Wiley & Sons]
[12] Raghavan V 2006 *J. Phase Equil. Diffusion* **27** 372

[13] Kopeček J, Sedláková-Ignácová S, Jurek K, Jarošová M, Drahokoupil J, Šittner P and Novák V 2009 *Structure development in Co38Ni33Al29 ferromagnetic shape memory alloy.* in: ESOMAT 2009 - 8th Eur. Symp. Martensitic Transformations. (Šittner P, Paidar V, Heller L and Seiner H; Eds.) 02013 [https://www.esomat.org/articles/esomat/abs/2009/01/contents/contents.html]
[14] Kopeček J, Jurek K, Jarošová M, Drahokoupil J, Sedláková-Ignácová S, Šittner P and Novák V 2010 *IOP Conf. Series: Mat. Sci. Eng.* **7** 012013
[15] Däbritz S, Langer E and Hauffe W 1999 *J. Anal. At. Spectrom.* **14** 409-412
[16] Däbritz S, Langer E and Hauffe W 2001 *Appl. Surf. Sci.* **179** 38-44
[17] Langer E, Däbritz S, Schurig C and HauffeW 2001 *Appl. Surf. Sci.* **179** 45-48
[18] Langer E, Kurt R and Däbritz S 1999 *Cryst. Res. Technol.* **34** 801-816
[19] Hubert-Protopescu M and Hubert H 1991 *Aluminium-cobalt-nickel ternary alloys: A comprehensive compendium of evaluated constitutional data and phase diagrams. Vol. 4 Al-Cd-Ce to Al-Cu-Ru.* (Petzow G and Effenberg G; Eds.) [Weinheim, Germany: VCH] 234
[20] Grün A, Henig E-T and Sommer F 1998 *Z. Metallkd.* **89** 591
[21] Breuer J, Sommer F and Mittemeijer E J 2001 *Metall. Mater. Trans. A* **32** 2157

BadgerFilm: A new thin film analysis programme (with other useful applications for EPMA)

A Moy and J H Fournelle

University of Wisconsin, Department of Geoscience, 53706 Madison, Wisconsin, U.S.A.

E-mail: amoy6@wisc.edu

Abstract. Characterisation of thin films upon substrates, i.e., thickness and composition determination, can be performed using the non-destructive technique of electron probe microanalysis (EPMA). This technique, first introduced in the 1960's and refined in the 1990's, is based on the accurate description of the electron ionisation depth distribution, the so-called $\phi(\rho z)$-distribution. Using these models, several thin film analysis computer programmes have been developed and distributed. However, these programmes have some limitations: (1) older programmes may not be compatible with modern computer systems; (2) the majority of the programmes are "black boxes" making it difficult to assess the reliability of the quantification results; and (3) commercial programmes can be expensive and this might prevent a lab from acquiring one of them especially when demands for thin film quantification are occasional. And experience has shown that where an EPMA lab is searching for funds and new users/customers, the ability to do thin film characterisation can be valuable. In this article, we present the thin film analysis programme BADGERFILM. The programme is free, open-source and implements documented $\phi(\rho z)$-models and algorithms. The programme is versatile and can be used to characterise multilayer samples but also bulk samples. BADGERFILM demonstrated excellent quantification results for the films and for the substrates when compared to other EPMA thin film analysis programmes and to other techniques. BADGERFILM can also be employed to calculate mass absorption coefficients from multi-voltage EPMA measurements and can calculate absolute X-ray intensities directly comparable to Monte Carlo simulations.

1. Introduction

Thin film materials are vital components in modern technologies, used in optical coatings, semiconductor devices, magnetic storage, superconductor research, etc. The direct characterisation of sub-micrometre to nanometre films requires specialised equipment and techniques [1]. The use of the electron microprobe for the analysis of thin films was soon recognised after the development of the technique by Raimond Castaing [2, 3]. Initially, the electron microprobe technique was not designed for thin film analysis due to early matrix correction methods assuming bulk homogeneous volumes. However, advancements in the $\phi(\rho z)$-distribution, improving matrix corrections in the 1990s, enabled accurate thin film characterisation. The technique is now employed in a wide variety of fields to determine, for example, coating thicknesses, the presence of oxidation layers or thin film homogeneity [4]. The $\phi(\rho z)$-distribution is also used for composition analysis of bulk homogeneous specimens. Multiple $\phi(\rho z)$-models have been developed [5], including the PAP, XPP, and XPHI models [6-8]. Traditional EPMA methods cannot characterise thin films (a single film on a substrate or a multilayer specimen) effectively; instead, specialised techniques like the multi-voltage analysis method are required. This involves measuring k-ratios at different voltages and comparing them to theoretical values obtained using an appropriate $\phi(\rho z)$-model in which the thickness and/or the composition of the layers or substrate are the unknowns. By successive iterations on these values, theoretical k-ratios are calculated until their values closely match the experimental k-ratios. This analysis procedure requires dedicated thin film analysis programmes. Commercial thin film analysis programmes are costly, discouraging some laboratories from acquiring them, except for the free but outdated GMRFILM programme developed by R. Waldo at General Motors Research in the late 1980's [9]. GMRFILM, being a DOS programme, is becoming difficult to run on modern computer. In addition, it is limited to single voltage analysis and loses the benefits of multi-voltage analysis. Also, most thin film analysis programmes are "black boxes," lacking transparency in their algorithms. To address the need for free and transparent thin film analysis software, we have developed BADGERFILM, an open-source EPMA thin film analysis programme. BADGERFILM implements widely used $\phi(\rho z)$-models (PAP, XPP, and XPHI) and utilises a non-linear fitting procedure (Levenberg-Marquardt algorithm [10, 11]) for convergence on k-ratios. It also accounts for secondary fluorescence and implements recent atomic parameters for absolute X-ray intensity calculations. BADGERFILM is written in .NET Visual Basic, is open source and can be freely downloaded at: https://github.com/Aurelien354/BadgerFilm.

2. X-ray generation

To analyse thin films by EPMA, an accurate description of the primary electron ionisation distribution is necessary. The $\phi(\rho z)$-distribution is used to calculate the primary X-ray intensities generated in the sample and emitted outside the sample. The distribution is also used to calculate the secondary fluorescence generated by the primary characteristic X-rays. Finally, this distribution can also be adapted to calculate the secondary fluorescence created by the bremsstrahlung.

2.1. *Primary characteristic X-rays*

In the general case of a layer buried in a multilayered sample, the primary X-ray intensity I_i^s emitted by an element i in the layer s located between mass depth ρz_s and ρz_{s+1} can be expressed by:

$$I_i^s = n_{\text{el}} \frac{N_{\text{A}}}{A_i} C_i \, \omega_j \, p_i \, \lambda_j \, \sigma_j(E_0) \, T_i^s \int_{\rho z_s}^{\rho z_{s+1}} \phi_j(\rho z) \, e^{-\chi_i^s (\rho z - \rho z_s)} \, d\rho z \; \varepsilon \, \frac{\Omega}{4\pi}, \quad (1)$$

where n_{el} is the number of incident electrons of energy E_0 reaching the sample per second. N_{A} is Avogadro's number, A_i represent the atomic weight of element i and C_i the weight fraction of element i. Then follow the relaxation parameters. The atomic parameter ω_j describes the fluorescence yield of shell (or subshell) j. The term p_i is the line fraction and is equals to $\Gamma_{j-k}/\Gamma_{j-total}$ where Γ_{j-k} represents the radiative transition probability for an electron to make a transition from shell k to shell j and $\Gamma_{j-total}$ represents the total radiative width for all possible transitions to the j shell. To take into account the fact that electron vacancies in the shell of interest can be created by migration of vacancies between subshells of the same shell through non-radiative transitions (Coster-Kronig and super-Coster-Kronig transitions), and between subshells of different, most inner, shells through radiative and non-radiative transitions, the enhancement factor λ_j is introduced. This factor is also sometimes denoted $1 + T_{CK}$.$\sigma_j(E_0)$ is the ionisation cross-section of the shell (or subshell) j by electron impact of incident energy E_0. The detector characteristics are represented by the terms ε and Ω/4π which are the intrinsic detection efficiency and the solid angle of collection of the detector, respectively. Note that when calculating the k-ratio, i.e., the ratio of the X-ray intensities measured on the unknown and on a standard under the same instrumental conditions, the electron impact ionisation cross-sections $\sigma_j(E_0)$ present outside of the integral (Eq. 1), as well as the relaxation parameters and the detector efficiency terms, cancel out between the numerator and the denominator.

The term T_i^s describes the absorption of the emitted characteristic X-rays by the l layers above the layer s and is given by:

$$T_i^s = \prod_{l=1}^{s-1} e^{-\chi_i^l (\rho z_l - \rho z_{l-1})}, \qquad (2)$$

where χ_i^l is the reduced mass absorption coefficient:

$$\chi_i^l = \left(\frac{\mu}{\rho}\right)_i^l \frac{1}{\sin\theta}, \qquad (3)$$

with $\left(\frac{\mu}{\rho}\right)_i^l$ being the mass absorption coefficient (MAC) of the material constituting the layer l for the relevant radiation produced by the element i. The term θ represent the takeoff angle of the detector. The term χ_i^s in Eq. (1) is the reduced MAC for the layer s. In Eq. (2), the term $\rho z_l - \rho z_{l-1}$ corresponds to the mass thickness of layer l, hence $e^{-\chi_i^l (\rho z - \rho z_s)}$ represents the absorption of primary characteristic X-rays in the layer l. In Eq. (1), $\phi_j(\rho z)$ represents the ionisation distribution of the shell j of element i. The integral term is calculated analytically or numerically, depending on the $\phi_j(\rho z)$-model chosen, over the mass thickness of the layer (or substrate) s to obtain the total emitted X-ray intensity.

However, $\phi(\rho z)$-distribution models must be adapted in the case of layered specimens. Indeed, as seen in Figure 1, the ionisation depth distribution in a thin film on top of a substrate is different from the distribution in a bulk homogeneous specimen of the same material because of perturbating effects, such as the backscattered electrons from one layer to another. When the atomic number of the substrate increases, the number of electrons backscattered into the film increases, leading to an increase of the $\phi(\rho z)$-distribution in the film. In the following, we briefly describe two $\phi(\rho z)$-models implemented in BADGERFILM, as well as their modifications to handle the case of layered specimens.

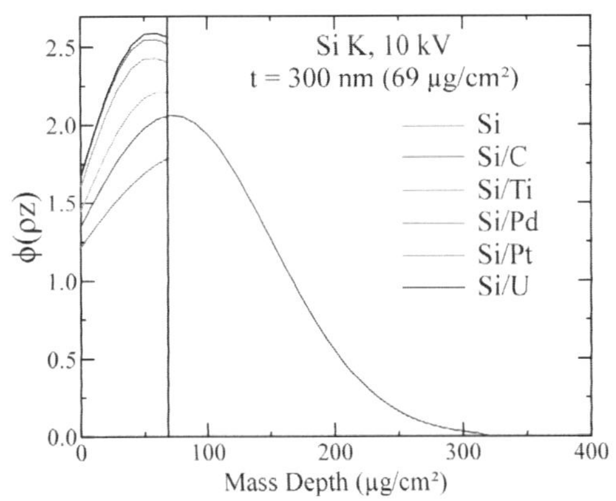

Figure 1. Effects of the substrate material on the $\phi(\rho z)$-distribution of the Si-K shell at 10 keV in a 300 nm thick film of Si. The distribution in pure Si is also shown. Results obtained with BADGERFILM using the XPHI model [8].

2.2. *The PAP model*

The PAP model was developed by Pouchou and Pichoir at the end of the 1980's [12,13] and refined in the 1990's [6]. The model is based on the definition of the $\phi(\rho z)$-distribution given by Castaing [14] stating that the area of the $\phi(\rho z)$-distribution is proportional to the number of primary ionisations in the shell j of element i, N_j, produced per incident electron:

$$N_j = \frac{C_i}{A_i} N_A \, \sigma_j(E_0) \int_0^\infty \phi_j(\rho) \, d\rho z \, . \qquad (4)$$

Pouchou and Pichoir assumed that the number of primary ionisations can be obtained irrespective of the modelling process, using the so-called continuous electron deceleration approximation:

$$N_j = C_i \frac{N_A}{A_i} R \frac{1}{S}, \quad (5)$$

where R is the backscatter loss factor and *1/S* is the stopping factor. Hence, we can write:

$$R \frac{1}{S} \frac{1}{\sigma_j(E_0)} = \int_0^{\infty} \phi(\rho z)\, d\rho z\,. \quad (6)$$

The expressions to calculate the factors R, *1/S* and $\sigma_j(E_0)$ are given elsewhere [5, 6, 15]. Eq. (6) imposes a condition on the area of the $\phi(\rho z)$-distribution, but not on its form. Pouchou and Pichoir choose to describe the ionisation depth distribution along the target mass depth using two connected parabolas, as shown in Figure 2:

$$\phi(\rho z) = \begin{cases} A_1(\rho z - R_m)^2 + B_1 & for\ 0 < \rho z \le R_c \\ A_2(\rho z - R_x)^2 & for\ R_c < \rho z < R_x \end{cases}, \quad (7)$$

where ρz represents the mass depth, R_m, R_c and R_x are the mass depth at which the distribution reaches its maximum, the mass depth at which the two parabolas are connected, and the maximum mass depth at which the ionisation ceases, respectively. By imposing the continuity of $\phi(\rho z)$ in R_c as well as the continuity of $\phi'(\rho z)$ (the derivative of $\phi(\rho z)$) also in R_c, the parameters in Eq. (7) can be determined:

$$A_1 = \frac{\phi(0)}{\left(R_m\left(R_c - R_x\left(\frac{R_c}{R_m} - 1\right)\right)\right)}, \qquad B_1 = \phi(0) - A_1 {R_m}^2, \qquad A_2 = A_1 \frac{R_c - R_m}{R_c - R_x} \quad (8)$$

where $\phi(0)$ is the surface ionisation, i.e., the value of the $\phi(\rho z)$-distribution at the surface of the sample. The expressions to calculate A_1, A_2, B_1 and $\phi(0)$ are given elsewhere [5, 6, 15].

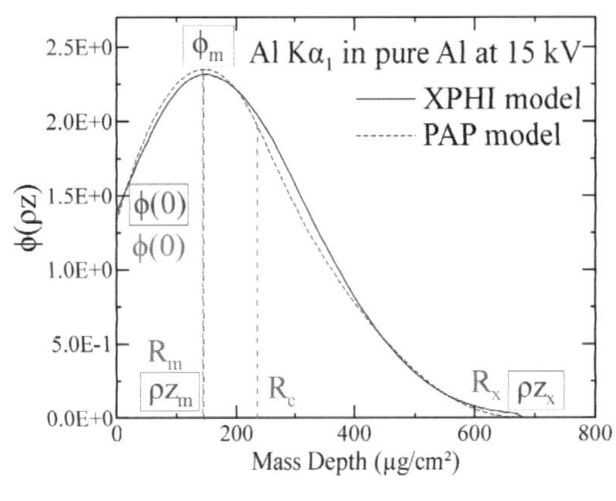

Figure 2. Si-K ionisation depth distribution as a function of the mass depth calculated using the PAP and the XPHI models at 15 kV. The parameters of the models are also indicated.

To adapt the PAP model to the case of a layered specimen, a weighting equation is employed to generate a set of fictitious bulk compositions with which the form parameters R_m, R_c, R_x and $\phi(0)$ are calculated. The weighting law used is a fourth order polynomial of the form [6,16]:

$$p(\rho z, L_r, R_r) = N(\rho z - L_r)^2(\rho z - R_r)^2\,, \quad (9)$$

where L_r and R_r are the roots of Eq. (9) and are restricted to the interval [$-R_x$; R_x]. To constrain the area under p to be equal to 1 between 0 and R_x, the normalisation factor N is employed. To each form parameter corresponds different values of L_r and R_r to give more or less weight to the first layer(s) of the stratified specimen and hence to give a better approximation of the associated form parameter. Using the appropriate values of L_r and R_r, the fictitious concentrations C_i^f of the i elements composing a given layer (or the substrate) are calculated by:

$$C_i^f = C_i \times N \times \int_{\rho z_1}^{Min(\rho z_2;R_x)} p(\rho z, L_r, R_r)\, d\rho z \,, \tag{10}$$

where ρz_1 and ρz_2 are the limits of the considered layer (or substrate) and ρz_2 must not exceed the maximum mass depth ionisation R_x. Then, all the fictitious compositions, from the layers and the substrate, are summed and normalised to 100 wt% to create a fictitious bulk composition used to calculate the different parameters of the $\phi(\rho z)$-distribution. R_x is calculated in an iteration loop with $R_r = R_x$ and $L_r = -0.4R_x$. To give more weight to the layers close to the surface, used for example to calculate $\phi(0)$, the roots $R_r = 0.5R_x$ and $L_r = -0.4R_r$ are used. $R_r = 0.7R_x$ and $L_r = -0.6R_r$ are used to calculate the deceleration factor *1/S* used in the expression of R_c. R_m is calculated using an average of the mean atomic numbers obtained during the calculations of $\phi(0)$ and R_c. See [6] or [16] for more details.

2.3. The XPHI model

The XPHI model was developed in the early 1990's by Merlet [17]. While the PAP model uses two connected parabolas to describe the $\phi(\rho z)$-distribution, the XPHI model uses a double-Gaussian distribution:

$$\phi(\rho z) = \begin{cases} \phi_m\, e^{-\frac{(\rho z-\rho z_m)^2}{\beta^2}} & for\ 0 < \rho z \le \rho z_m \\ \phi_m\, e^{-\frac{(\rho z-\rho z_m)^2}{\alpha^2}} & for\ \rho z_m < \rho z \le \rho z_x \end{cases} , \tag{11}$$

with $\alpha = 0.46598\,(\rho z_x - \rho z_m)$ and $\beta = \frac{\rho z_m}{\sqrt{ln(\phi_m/\phi(0))}}$. As shown in Figure 2, the XPHI model uses four form factors: $\phi(0)$ the surface ionisation, ϕ_m the maximum of the $\phi(\rho z)$-function, ρz_m the mass depth at which ϕ_m occurs, and ρz_x the mass depth at which the function becomes negligible ($\phi(\rho z_x) \le 0.001$). The different expressions to calculate these form parameters are given elsewhere [5, 18]. The advantage of a description using Gaussian functions is that Eq. (1) can easily be integrated analytically:

$$I_i^s = cst\; n_{\text{el}}\, C_i\, T_i^s \frac{\sqrt{\pi}}{2} \phi_m \left\{\beta e^{-\gamma}\left[erf\left(\frac{\delta}{\beta}\right) + erf\left(\frac{\chi\beta}{2}\right)\right] + \alpha e^{-\zeta}\left[1 - erf\left(\frac{\chi\alpha}{2}\right)\right]\right\} , \tag{12}$$

with $cst = \varepsilon\, \frac{\Omega}{4\pi} \frac{N_A}{A_i}\, \omega_j\, p_i\, \lambda_j\, \sigma_j(E_0)$, $\gamma = \rho z_m \chi - \left(\frac{\chi\beta}{2}\right)^2$, $\delta = \rho z_m - \frac{\chi\beta^2}{2}$, and $\zeta = \rho z_m \chi - \left(\frac{\chi\alpha}{2}\right)^2$. Erf(x) denotes the error function.

As described previously, for the case of multilayer specimens, the $\phi(\rho z)$-distribution must be modified. The XPHI model, as implemented in XFILM [8], uses the weighting procedure originally introduced in [19] to calculate backscattering coefficients on thin films. The four form parameters are calculated using a hyperbolic tangent weighting function:

$$f_{F/S} = (f_F - f_S)\tanh(Ax + Bx^2) + f_S \,, \tag{13}$$

where f denotes either of the form parameters. The subscript F/S denotes the form parameters calculated for the film over substrate specimen, and the subscripts F and S denote the form parameters calculated in a bulk sample with a composition similar to the film or substrate, respectively. The parameters A and B are given by:

$$A = \frac{2.153\,Z_F - 14.798}{3.706\,Z_F + 17.822} \quad , \quad B = \frac{0.3618\,Z_F - 27.80}{1.235\,Z_F - 82.055}, \tag{14}$$

where Z_F is the mean atomic number of the film: $Z_F = \sum_i C_i Z_i$ (with C_i and Z_i the concentration and the atomic number of the element i, respectively). The parameter x is a reduced thickness, function of the mass thickness t of the film and given by $x = 4t/(\rho z_x)$. The weighting function given by Eq. (13) ensures that for very thick films, the calculated form parameters are close to f_F while for very thin films, the values tend toward f_S.

In the case of a multilayer specimen composed of N layers on top of a substrate (the layer just on top of the substrate having the index $i = N$ and the layer at the surface of the sample having the index $i = 1$), the form parameters for the layer i are given by Eq. (13) in which f_F is replaced by the form factor of the layer i, f_i, and in which f_S is replaced by the form factor of the layer $i-1$, f_{i-1}, i.e., the layer beneath the layer i. Practically, the form factors of the layer N directly on top of the substrate must be calculated first using Eq. (13), then the form factors of the layer $N-1$ can be calculated, etc. This process is repeated until the form factors of the layer at the surface (index $i = 1$) can be calculated. The parameters A, B and x are also calculated for each layer using the current layer's (layer i) composition and mass thickness.

2.4. *Characteristic fluorescence*

The characteristic fluorescence is defined as the production of characteristic X-rays by other characteristic X-rays of higher energy. In some situations, the characteristic fluorescence can amount up to 30 % of the total X-ray intensity and should not be neglected. A classic example is the fluorescence of the Fe-Kα X-ray line by the Ni Kα- and Kβ-lines. At 25 kV, the Fe-Kα X-ray intensity of a 100 nm film of $FeSi_2$ deposited on a Ni-substrate is increased by 9.5 % due to the characteristic fluorescence by Ni. The characteristic fluorescence is generally calculated numerically by considering all the characteristic X-rays with an energy higher than electron shell energy giving rise to the X-ray line of interest. These exiting characteristic X-rays come from all the other elements present in the electron interaction volume but also from the same element producing the studied X-rays, e.g., the Fe-Lα X-rays can be excited by the Fe-Kα X-rays if these latter X-rays are produced. For each fluorescing X-ray line, the associated $\phi(\rho z)$-distribution is calculated, and the X-ray intensity calculated as a function of the mass depth. This latter function is then integrated along the mass depth of the relevant layer to calculate the fluorescence X-ray intensity. More details on the calculations can be found in [16].

2.5. *Bremsstrahlung fluorescence*

The bremsstrahlung fluorescence can amount up to a few percent of the total produced characteristic X-ray intensity. This effect increases as the difference between the measured characteristic X-ray and the electron beam energy increases because more bremsstrahlung photons are available to excite the characteristic line of interest. This effect can be particularly important in thin films. A simple model to calculate the bremsstrahlung fluorescence is to assume that the bremsstrahlung X-rays are produced at a point on the surface of the considered layer [6]. A more refined model, as implemented in BADGERFILM, consists in calculating a $\phi(\rho z)$-distribution representative of the bremsstrahlung distribution of energy E_{ph}. To the best of the authors knowledge, there is no parametric model describing the bremsstrahlung distribution as a function of the bremsstrahlung energy E_{ph}. However, it is possible to approximate this distribution by using the PAP or XPHI $\phi(\rho z)$-models and by considering a fictitious element having a characteristic X-ray with a critical ionisation energy E_{ph} equals to the energy of the considered bremsstrahlung photon. This distribution denoted $\phi_{Brem}(\rho z; E_{ph})$ represents the distribution of bremsstrahlung X-rays of energy E_{ph} as a function of the mass depth ρz for a given material and electron beam energy. These $\phi(\rho z)$-distributions can then be integrated over the depth of the specimen to calculate fluorescence, following the same method as for the characteristic fluorescence. All the X-rays energies, from the electron beam energy E_0 down to the electron ionisation threshold of the characteristic X-ray of interest, E_c, as well as their associated ϕ_{Brem} distribution should be considered. In practice, the energy range $[E_c; E_0]$ is discretised into a finite subset of energies at which the ϕ_{Brem} distributions are calculated. This discretisation should also include the energy of the absorption edges of the elements constituting the specimen because the bremsstrahlung X-ray spectrum (as a function of the photon energy) presents discontinuities at these energies. In each energy interval, the total emitted bremsstrahlung fluorescence intensity is obtained by numerical integration.

However, the ϕ_{Brem} distributions obtained using the PAP or XPHI models are only an approximation of the real bremsstrahlung distribution. To help improve the model, the shape of the calculated X-ray spectrum (as a function of the photon energy) produced only by bremsstrahlung fluorescence is weighted by the shape of the bremsstrahlung spectrum $I(E)$ as proposed by Small *et al.* [20] and refined by Kulenkampff (as cited in [6]):

$$I(E) = q\, e^{B} \left[\bar{Z}\left(\frac{E_0}{E} - 1\right)\right]^{M} + C\,\bar{Z}^2, \tag{15}$$

where q is a scaling factor of 10^{-8} keV^{-1} [6], $B = -3.22 \times 10^{-2} \times E_0 + 5.8$, $M = 5.99 \times 10^{-3} \times E_0 + 1.05$ and $C = 6 \times 10^{-10}$. The mean atomic number of the material $\bar{Z}$ is given by $\bar{Z} = \sum_i C_i Z_i$, where C_i and Z_i are quantities previously defined.

This method, despite producing more accurate results than the point surface method, appears to overestimate the bremsstrahlung fluorescence intensity when compared to Monte Carlo simulations.

To improve the calculation of the bremsstrahlung fluorescence, BADGERFILM implements two additional factors in Eq. (15):

$$I(E) = \beta(\bar{Z})q\, e^{B} \left[\bar{Z}\left(\frac{\alpha(\bar{Z})\,E_0}{E} - 1\right)\right]^{M} + C\,\bar{Z}^2, \tag{16}$$

where $\alpha(\bar{Z})$ modifies the energy of the primary electrons as a function of the mean atomic number of the material. This parameter is also introduced in the calculation of the bremsstrahlung ϕ_{Brem} distribution and modifies the primary electron beam energy in a similar way. $\beta(\bar{Z})$ is a factor that modifies the value of the constant q. These parameters were obtained by adjusting their values until the calculated bremsstrahlung fluorescence intensities match bremsstrahlung fluorescence data obtained by Monte Carlo simulations using the code PENEPMA [21]. This fitting procedure was performed on pure elements using the main Kα, Lα, and Mα characteristic X-ray lines. The obtained values of $\alpha(\bar{Z})$ and $\beta(\bar{Z})$ can easily be fitted by part using polynomial functions, as seen in Figure 3 for the Kα-line, with good coefficients of determination R^2. The obtained polynomials are implemented in BADGERFILM to calculate the bremsstrahlung fluorescence.

3. BadgerFilm

BADGERFILM is an open-source software with a user-friendly graphical interface designed to facilitate the input and processing of data for thickness and composition calculations of multilayered specimens. The previously described algorithms were implemented in BADGERFILM. The programme can also be used for traditional quantification of bulk homogeneous samples and to determine MAC values using the multi-voltages method [22]. The programme offers flexibility in selecting parameters such as the $\phi(\rho z)$-model, the MAC model, or ionisation cross-section model for quantification. Data required for quantification includes the element symbol, characteristic X-ray line, experimental *k*-ratio, accelerating voltage, and optionally, the error on the *k*-ratio. The standard on which the *k*-ratio has been measured, and previously set up in BADGERFILM, needs to be specified. BADGERFILM supports various

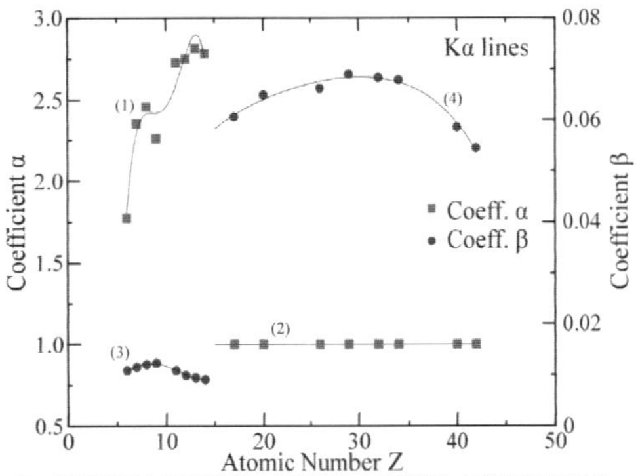

(1) y=-3.737141E-3x⁴ + 1.519312E-1x³ - 2.267005x² + 1.478875E1x - 3.329064E1; R²=0.933
(2) y=1; R²=1
(3) y=4.476461E-6x⁴ - 1.535630E-4x³ + 1.743002E-3x² - 7.318512E-3x + 1.937490E-2; R²=0.991
(4) y=-1.153359E-7x⁴ + 1.138019E-5x³ - 4.570208E-4x² + 9.165271E-3x - 9.048994E-3; R²=0.977

Figure 3. α and β coefficient values used in the calculation the bremsstrahlung fluorescence for the Kα X-ray lines. Symbols are data obtained by fitting the coefficients to Monte Carlo simulations. Continuous lines represent the polynomial fits of the coefficients over the atomic number Z.

characteristic X-ray lines, both traditional and less traditional ones: Kα (or $K\alpha_1$, or $K\alpha_2$), Kβ, Lα (or $L\alpha_1$, or $L\alpha_2$), $L\beta_1$, $L\gamma_1$, Lη, Lℓ, L_3N_5, Mα (or $M\alpha_1$, or $M\alpha_2$), $M\beta_1$, $M\gamma_1$, M_4O_6, M_3O_5, M_2N_4, M_2N_1, M_3O_1, M_3O_4, M_1N_2, M_1N_3, M_2O_4 and M_1O_2. To fit the data and converge towards a solution, the programme uses the Levenberg-Marquardt non-linear fitting method [10, 11], varying film compositions and thicknesses until calculated *k*-ratios match experimental values. By default, BADGERFILM uses the same atomic parameters as those employed by the Monte Carlo code PENELOPE 2018 [23] and its EPMA module PENEPMA [21]. The default MACs are derived from the PENELOPE database which were calculated from the photoelectric cross section calculated by Sabbatucci and Salvat [24]. The default electron impact ionisation cross-sections are those of Bote and Salvat [25] and the atomic transition rates (used to calculate X-ray production cross sections) are from the LLNL Evaluated Atomic Data Library [26]. It should be noted that the electron impact ionisation cross-section models originally used in the calculation of the $\phi(\rho z)$-models are unchanged as they are intrinsically related to the good performance of the model. These parameters are only used to calculate absolute X-ray intensities or to determine MACs. Users have the option to select other atomic parameter models and provide experimental MAC values for specific emitters, X-ray lines, and absorbers. The calculated compositions and film thicknesses are accompanied by an uncertainty estimate derived from the fitting algorithm and *k*-ratio uncertainty. If the user does not provide a *k*-ratio uncertainty, a 5 % error is assumed. Uncertainties related to other variables and matrix correction models are not considered. BADGERFILM allows plotting of calculated and experimental *k*-ratios against the accelerating voltage for easy evaluation of the fit quality. Data can be saved in a BADGERFILM text file for future use. STRATAGEM files (version 6.2) can also be imported into BADGERFILM, enabling compatibility with STRATAGEM [27] and Probe for EPMA [28]. Calculated results can be conveniently exported to a spreadsheet programme in a pre-formatted layout.

4. Thin film analysis

BADGERFILM is an EPMA software dedicated to the determination of thin film thicknesses and composition. It uses the multi-voltage method where experimental *k*-ratios are measured at different accelerating voltages. We present below two test examples performed on *i*) a thin film on a substrate, and *ii*) a multilayer specimen.

4.1. Thin film on substrate

The simple case of the thickness determination of a C-film deposited on a Si-substrate is shown. Experimental C-Kα and Si-Kα *k*-ratios (relative to pure standards) were acquired by Llovet and Merlet [8] at multiple accelerating voltages ranging from 2 to 25 kV. Using the XPHI model as implemented in XFILM, these authors estimated the C-film thickness to be 137 nm (assuming a C density of

IOP Conf. Series: Materials Science and Engineering **1324** (2025) 012010 doi:10.1088/1757-899X/1324/1/012010

2.2 g/cm^3). Using the PAP model and the PENELOPE MACs, BADGERFILM estimated the C-film thickness to be 135.1 ± 1.7 nm, similar to the value obtained by Llovet and Merlet. Figure 4 shows the excellent agreement between the experimental *k*-ratios and the BADGERFILM calculated *k*-ratios.

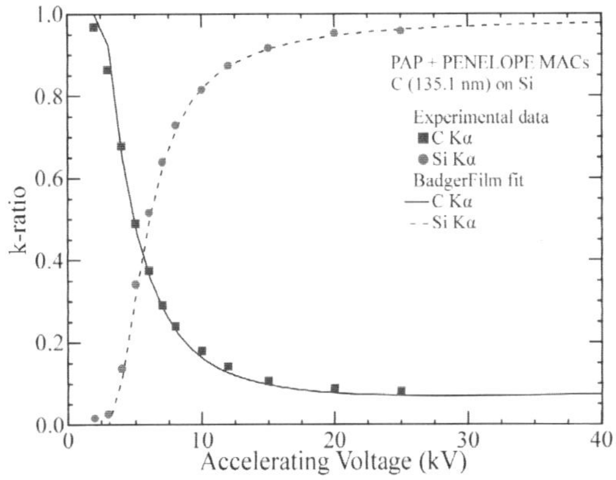

Figure 4. Determination of the thickness of a C-film deposited on a Si-substrate using the C- and Si-Kα *k*-ratios. Symbols are experimental *k*-ratios from [8]. Continuous and dashed lines are *k*-ratios calculated by BADGERFILM.

4.2. Multilayer specimens

BADGERFILM iterates on the composition and thickness of the films (and substrate) to calculate theoretical *k*-ratios until they match the experimental *k*-ratios. Multilayered specimen can easily be analysed using this technique. As shown in Figure 5, the compositions and thicknesses of a bilayer of Ni-Cr/Fe-Gd-Pt on top of a Si substrate is determined using three accelerating voltages at 20, 25 and 30 kV. Experimental data acquired by Pouchou were extracted from [29]. Experimental *k*-ratios (relative to pure standards) for the Ni-Kα, Cr-Kα, Fe-Kα, Gd-Lα and Pt-Mα were processed by BADGERFILM using the PAP model and MACs extracted from the PENELOPE database. The obtained results are as follow: a 68.3 nm thick first layer with a composition of 14.4 wt% Ni and 85.6 wt% Cr, and a 24.0 nm thick second layer with a composition of 50.3 wt% Fe, 29.4 wt% Gd and 20.3 wt% Pt. These results are in excellent agreement with the original results of Pouchou obtained using STRATA (the ancestor of STRATAGEM), as well as with results obtained with other thin film analysis programmes and other thin film characterisation techniques, such as the Rutherford Backscattering technique [16]. PENEPMA Monte Carlo simulations were also performed using the films thickness and composition found by BADGERFILM. Simulated *k*-ratios are in excellent agreement with the experimental *k*-ratios and the BADGERFILM calculated *k*-ratios (Figure 5), supporting the good accuracy of the method.

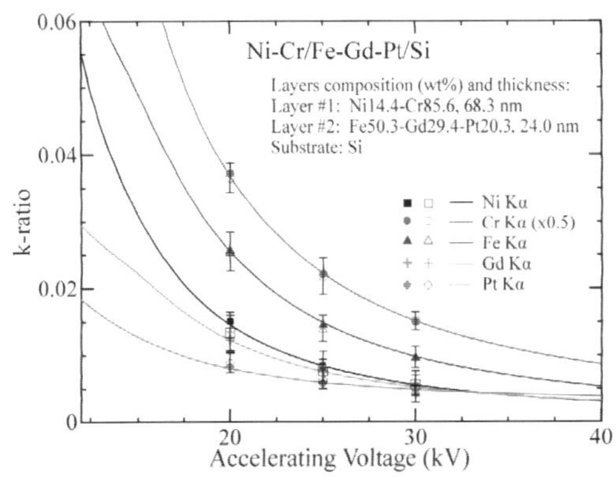

Figure 5. Compositions and thicknesses of a multilayer specimen determined using *k*-ratios. Solid symbols are experimental *k*-ratios from [8]. Continuous lines are calculated by BADGERFILM. Open symbols are from PENEPMA simulations (error bars correspond to 3-sigma uncertainty).

5. Bulk specimens

BADGERFILM is not only dedicated to the analysis of thin film samples but can also accurately quantify bulk specimens [15]. Oxygen is not always routinely measured, especially in geological materials, mainly because of spectral interferences (notably with EDS spectrometers), because of the difficulty of having a well characterised O-standard (O is strongly absorbed by the coating of the standard and any uncertainty in the coating thickness will result in uncertainties in the quantification results) or simply to reduce the analysis time. In such situation, O is determined by stoichiometry relative to the other cations. BADGERFILM offers the possibility to quantify samples using O defined by stoichiometry as well as another element defined by stoichiometry relative to O. As an example, the quantification of a dolomite specimen, (ideal stoichiometric formula $CaMg(CO_3)_2$), can be done by only measuring Ca and Mg (and trace elements such as Fe and Sr if present) and by determining O by stoichiometry and C relative to O using a 1:3 atomic ratio. The correct appropriation of O and C is made during the matrix correction iterations, rather than after the calculations. This allows these elements to be fully accounted for in the matrix correction procedure and avoid any quantification bias that can result when these elements are added after the correction (Table 1).

Table 1. Elemental concentration (in wt. fraction) of a dolomite sample with O and C either determined by stoichiometry during the matrix correction procedure or not considered.

	Fe	Mn	Ca	Mg	O	C	Total
With O, C	0.0093	0.0014	0.2117	0.1252	0.5104	0.1277	0.9857
No O, C	0.0098	0.0015	0.2210	0.1222	-	-	0.3545
Error %	5.4	7.1	4.4	-2.4			

6. MAC determination

BADGERFILM can also be used to determine MACs by using multi-voltage measurements on specimens of known composition and geometry (film thickness). It treats the MAC for a specific emitter, characteristic X-ray line, and absorber as unknown, iterating until calculated and experimental *k*-ratios converge. To illustrate this technique, the MAC of the Si-Kα line by Hf in $HfSiO_4$ can be estimated by recording the Si-Kα X-ray intensity at several accelerating voltages. The data were processed with BADGERFILM using the PAP $\phi(\rho z)$-model as well as MACs extracted from the PENELOPE database for the MACs other than the one of interest. As shown in Figure 6, the MAC value found by BADGERFILM of 3,172 cm^2/g is in good agreement with the value found by Donovan [28] of 3,477 cm^2/g. In comparison, the theoretical MAC extracted from the FFAST [30] and MAC30 [31] databases show significant discrepancies from experimental MACs. These results suggest a need for a re-evaluation of some MACs, particularly for low-energy X-rays and high-*Z* absorbers (see Llovet *et al.* [32, 33]).

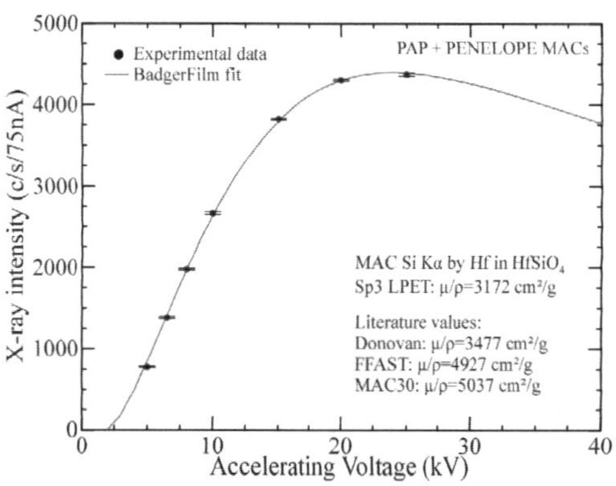

Figure 6. Determination of the Si-Kα by Hf MAC in $HfSiO_4$ by BADGERFILM. Symbols are experimental X-ray intensities. The continuous line is the result of the fitting by BADGERFILM using the PAP $\phi(\rho z)$-model and PENELOPE MACs. The result is compared to experimental [28] and theoretical [30,31] values form the literature.

7. Conclusion

BADGERFILM is a versatile, free and open-source EPMA analysis programme, for the quantification of multi-layered specimens, as well as bulk samples. The programme implements several $\phi(\rho z)$-models, MACs datasets, as well as atomic parameters. BADGERFILM was designed to be flexible and easily modifiable, to be adapted to new types of problems that EPMA users may encounter. The programme has been shown to give accurate thin film characterisation results when compared to other thin film programmes and other characterisation methods. BADGERFILM can be downloaded at the following address: https://github.com/Aurelien354/BadgerFilm

Acknowledgements

Support for this research came from the National Science Foundation: EAR-1337156 (JHF), EAR-1554269 (JHF) and EAR-1849386 (JHF).

References

[1] Llovet X, Moy A, Pinard P T and Fournelle J H 2020 *Prog. Mater. Sci.* **116** 100673
[2] Sweeney W E, Seebold R E and Birks L S 1960 *J. Appl. Phys.* **31** 1061-1064
[3] Hunger H-J 1988 *Scanning* **10** 65-72
[4] Dumelié N, Benhayoune H and Balossier G 2007 *J. Phys. D. Appl. Phys.* **40** 2124-2131
[5] Lavrent'ev Y G, Korolyuk V N and Usova L V 2004 *J. Anal. Chem.* **59** 600-616
[6] Pouchou J-L and Pichoir F 1991 *Quantitative analysis of homogeneous or stratified microvolumes applying the model "PAP".* in: Electron probe quantitation. (Heinrich K F J and Newbury D E; Eds.) [New York, NY:(Springer] 31-75
[7] Merlet C 1995 *A new quantitative procedure for stratified samples in EPMA.* in: Proc. 29th Ann. Conf. Microbeam Analysis Society. (Etz E S; Ed.) [Weinheim, Germany: VCH Publishers, Inc.] 203-204
[8] Llovet X and Merlet C 2010 *Microsc. Microanal.* **16** 21-32
[9] Waldo R A 1988 *An iteration procedure to calculate film compositions and thicknesses in electron-probe microanalysis.* in: Microbeam analysis. (Newbury D E; Ed.) [San Francisco, CA: San Francisco Press, Inc.] 310-314
[10] Levenberg K 1944 *Q. Appl. Math.* **2** 164-168
[11] Marquardt D 1963 *J. Soc. Ind. Appl. Math.* **11** 431-441
[12] Pouchou J and Pichoir F 1984 *J. Phys.* **45** 17-20
[13] Pouchou J and Pichoir F 1984 *J. Phys.* **45** 47-50
[14] Castaing R 1951 *Application des sondes electroniques a une méthode d'analyse ponctuelle chimique et cristallographique.* PhD thesis. [Paris, France:, Université de Paris]
[15] Moy A and Fournelle J 2021 *Microsc. Microanal.* **27** 266-283
[16] Moy A and Fournelle J 2021 *Microsc. Microanal.* **27** 284-296
[17] Merlet C 1992 *Quantitative electron probe microanalysis: New accurate Φ(ρz) description. in: Electron Microbeam Analysis.* (Boekestein A and Pavićević M K; Eds.) [Vienna, Austria: Springer] 107-115
[18] Merlet C 1994 *Mikrochim. Acta* **114–115** 363-376
[19] Hunger H-J and Rogaschewski S 1986 *Scanning* **8** 257-263
[20] Small J A, Leigh S D, Newbury D E and Myklebust R L 1987 *J. Appl. Phys.* **61** 459-469
[21] Llovet X and Salvat F 2017 *Microsc. Microanal.* **23** 634-646
[22] Pouchou J L and Pichoir F M A 1988 *Determination of mass absorption coefficients for soft X rays by use of the electron microprobe.* in: Microbeam analysis. (Newbury D E; Ed.) [San Francisco, CA: San Francisco Press, Inc.] 319-324
[23] Salvat F 2019 PENELOPE 2018: A code system for Monte Carlo simulation of electron and photon transport. [Barcelona, Spain: Nuclear Energy Agency]
[24] Sabbatucci L and Salvat F 2016 *Radiat. Phys. Chem.* **121** 122-140
[25] Bote D and Salvat F 2008 *Phys. Rev. A - At. Mol. Opt. Phys.* **77** 1-24

[26] Perkins S T, Cullen D E, Chen M H, Rathkopf J, Scofield J and Hubbell J H 1991 Tables and graphs of atomic subshell and relaxation data derived from the LLNL Evaluated Atomic Data Library (EADL), Z = 1--100. Technical Report. [Livermore, CA: Lawrence Livermore National Laboratory]

[27] STRATAGem version 6.2. [Guyancourt, France: SAMx]

[28] Donovan J J, Kremser D, Fournelle J and Goemann K 2021 Probe for EPMA v. 12.9.5 User's Guide and Reference. [Eugene, Oregon: Probe Software, Inc.]

[29] Pouchou J-L 1993 *Anal. Chim. Acta* **283** 81-97

[30] Chantler C T, Olsen K, Dragoset R A, Chang J, Kishore A R, Kotochigova S A and Zucker D S 2005 *X-ray form factor, attenuation and scattering tables. (version 2.1)*. [Gaithersburg, MA: National Institute for Standards and Technology]

[31] Heinrich K F J 1987 Mass absorption coefficients for electron probe microanalysis. in: Proc. ICXOM XI (Brown J D and Packwood R H; Eds.) 67-119

[32] Llovet X, Pöml P, Moy A and Fournelle J H 2023 *Soft X-ray EPMA: choosing the right MACs*. in: Book of Tutorials and Abstracts of the EMAS 2023 Workshop. (ISBN 978 90 8227 6961)

[33] Llovet X, Pöml P, Moy A and Fournelle J H 2023 *Microsc. Microanal.* **29** 540-551

IOP Conf. Series: Materials Science and Engineering **1324** (2025) 012011 doi:10.1088/1757-899X/1324/1/012011

A new compact electron detector with REELS, EPES and elastic peak imaging capabilities for SEM

P Staib

Williamsburg Scientific Instruments, Staib Instruments, 295 McLaws Circle, 23185 Williamsburg, Virginia, U.S.A.

E-mail: pstaib@wscientific.com

Abstract. A new energy analyser to be used as add-on in SEM with multiple operating modes is described. It uses a high energy resolution of the retarding field to combine high energy resolution with high sensitivity and small size in order to fit inside the microscope vacuum chamber. The electron optical transmission of retarding mesh is obtained from computer simulations of the electron trajectories and a general formula for calculating the energy resolution is deducted. The analyser can perform as secondary electron detector for scanning imaging with energy filtering capability. The elastic peak is fully energy resolved so that recoil energy losses of light elements are measured and mapped. REELS measurements are shown to illustrate the high dynamic range and sensitivity of the system. The analyser is also sensitive to photons when the analyser energy is set above the primary beam energy, thus repelling all electrons and the detector gain increases correspondingly. Cathodoluminescent sample areas are mapped and soft X-rays are detected increasing the detector sensitivity up to single photon detection.

1. Introduction

Electron energy loss spectroscopy (EELS) is a well-established method for characterisation of elements in materials in transmission electron microscopes (TEM) and a comprehensive description of the technique and its applications is given in reference [1]. This work presents a system for measuring energy distribution of electrons reflected from the surface. There are several basic differences between the transmission and the backscattering modes: First, in transmission mode the high energy primary electron beam passes through a thin layer of material of controlled thickness chosen to optimise the intensity of energy losses and low beam damping and scattering. This is not possible in backscattering mode, the energy distribution being related to the mean free path and escape depth of electrons near the surface, which is primarily a function of the beam energy. Second, in transmission the emerging electron beam forms a dense well collimated beam of small angular aperture. In contrast, the backscattered signal covers a full angle range and its intensity is much weaker owing to the low backscattering factor. Therefore, different analyser designs are required for each application, Reflection electron energy loss spectroscopy (REELS) needs both a wide angular solid angle for sensitivity and a high energy resolution. A new energy analyser is described that can perform like a secondary electron detector for scanning electron microscope (SEM) imaging but, having high energy resolution and sensitivity, allows the measurement of backscattered electron distributions and performs REELS with microscopes operating in SEM mode. The solution presented is based on the powerful energy filtering properties of retarding field optics that can combine high sensitivity with high energy resolution with a small physical size able to fit inside a SEM chamber.

Performances of the system are illustrated by examples of energy resolved SEM imaging with special emphasis to imaging parts of the fully resolved elastic peak to perform elastic peak electron spectroscopy (EPES). Energy losses owing to the recoil energy of atoms are measured and imaged in scanning mode showing the distribution of light elements on the surface. The sensitivity of REELS measurements allows the scanning imaging of major energy losses. The analyser can operate at vacuum up to 10^{-4} torr to perform in situ and operando analyses in material deposition systems or preparation chambers.

2. Description of the energy analyser

The analyser is retarding field type energy filter as previously patented and used for energy filtering of reflection high-energy electron diffraction (RHEED) diagrams in growth chambers [2], for Auger electron spectroscopy (AES) [3, 5] and electron backscatter diffraction (EBSD) [4, 5] but with improved energy resolution and sensitivity to operate at low primary beam current.

Figure 1 shows a schematic view of the new retarding field energy analyser. Electrons emitted from the surface of the sample reach the entrance aperture of the analyser and are transmitted through the input lenses used to collimate the electron flux into a beam of parallel trajectories. The electrons impinge at normal incidence angle onto the flat retarding mesh electrode and are energy filtered. Transmitted electrons are extracted by exit lenses and are collected by a fluorescent screen. The light from the screen is detected by a photomultiplier located outside the vacuum and used to amplify the signal.

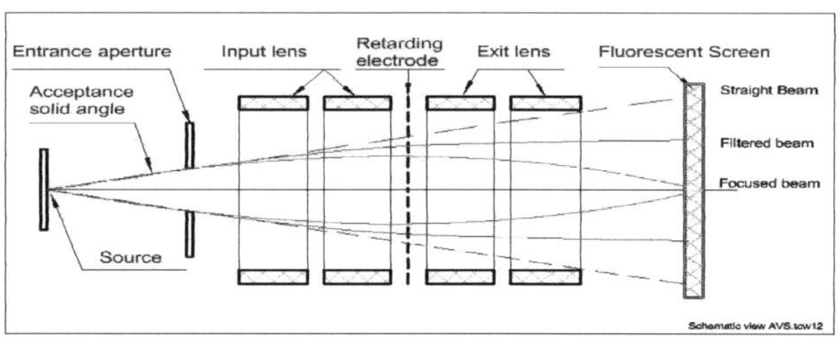

Figure 1. Schematic view of the energy analyser consisting of a retarding collimating lens, a retarding mesh and a re-focussing lens. Fluorescent screen is used as detector.

The entrance aperture defines the angular range to an acceptance solid angle Ω. The voltage of input lens unit is set to convert beam into a parallel beam for a specific beam kinetic energy. The beam reaches the flat conductive mesh with high transparency and small openings. Electrons of lower energy are repelled and those of higher energy are transmitted. The transmitted beam is extracted and focussed by the exit lens. The low energy part is focussed and reaches the centre of the screen, and electrons of higher kinetic energies are less focussed and reach the outer detector surface. The focussed centre spot always corresponds to electrons near the filter energy.

The transmitted signal is the integral distribution of the emitted $N(E)$ distribution of secondary electrons. When operating near the elastic peak the integrated signal remains low and the signal to noise ratio (S/N) is maximum, basically the same as to one of an energy bandpass analyser. The collected signal remains low in the characteristic energy loss region. It increases at lower energy thus reducing the S/N ratio of structures like core energy losses and Auger peaks located further away from the elastic peak.

The analyser has a full acceptance angle of 35°, a length sample to screen of 120 mm and a diameter of 85 mm. The working distance between sample and entrance aperture is 25 mm. The unit is inserted into the microscope vacuum chamber and mounted onto a CF6" OD vacuum window flange.

3. Detector system

The photomultiplier is attached outside the vacuum chamber, directly attached to the viewport. The photon signal from the fluorescent screen is collected by a photomultiplier amplifying the signal with a variable gain controlled by its voltage, which is located outside the vacuum chamber. The multiplier signal (PM signal) is used directly for scanning imaging and sent to a lock-in amplifier for spectroscopy. The lock-in amplifier performs the derivative of the direct photomultiplier (PM) signal when applying a modulation voltage to the retarding electrode and the output of the LockIn signal (LI signal) that restores the N(E) distribution of electrons. It is also possible to use the numerical derivative of the PM signal to retrieve the N(E) distribution. The LI signal is used for spectroscopy owing to the advantage of the noise reduction properties of the lock-in amplifier. The detector system is also sensitive to photons generated by cathodoluminescence, UV from plasmon decay and soft X-rays emission.

4. Transmission properties of energy filtering mesh electrodes

The mesh electrode is commonly made of a fine metallic mesh made of woven thin wires or of a thin metal sheet chemically etched. The size of the mesh apertures and the thickness of the material are the main parameters determining the transmission function and energy resolution. The spacing d between bars of a rectangular mesh is the main parameter. In the following calculations the full transmission at T = 100 % is normalised to the full area of the apertures. Simulations are performed for circular and rectangular aperture shapes.

The retarding electrode is polarised negatively to create a retarding field E_r in front of the mesh and an accelerating extraction field E_x in the back. The extraction field is required in order to guide and collect the transmitted electron flux. In order to perform scalable trajectory simulations, the fields are normalised to the size d of the mesh opening and is expressed in volts per distance d. As numerical example, a retarding voltage of 100 V applied to a mesh separated by 10 mm having 3 apertures per mm corresponds to a retarding field E_r = 3.33 V/d. The results can readily be scaled to the actual applied field values measured in V/d. The extraction field in the back of the mesh is expressed as the ratio of retarding and extraction fields E_r/E_x and play an important role in determining the energy resolution.

These two fields cause a field penetration inside the mesh apertures. The potential on the centre axis increases toward the mesh and reaches a maximum near the plane of the mesh followed by a decreases due to the extraction field. Position and value of the maximum depend on both fields, as does the transmission, and do not depend on the actual potential of the electrode.

For a thin electrode, the saddle point located at a position on the axis depends upon the field ratio E_x/E_r. For a symmetric field configuration the saddle point is located in the mesh plane. For lower extraction fields it moves into the extraction region. The potential difference between the hole centre and the mesh only depends on the values of the electrostatic fields applied in the front and in the back of the mesh electrode, and is independent of the actual voltage applied to the optical mesh assembly. More accurately, the voltage applied to the mesh sets the position of the Fermi level and the actual potential of the surface of the mesh is larger by the work function of the mesh material. This causes a constant offset of the retarding voltage that should be taken into account for absolute energy calibration.

A flux of collimated particles impinging the mesh at normal incidence at increasing kinetic energies is used to measure the transmission. The incoming flux is repelled until the kinetic energy reaches the retarding potential. First transmitted are the trajectories near the centre of the aperture. Progressively increasing energy, the outer trajectories are transmitted until reaching the edge of the aperture. There is a singularity concerning the edge transmission because the trajectories of incoming electrons are bent toward the lower potential of the centre and trajectories outside the edge are transmitted. The contribution is significant owing to the long length of the border edge.

Extensive simulations with variable ratio E_x/E_r lead to the approximate formula for the energy resolution measured at FWHM in meV

$$\Delta E = E_r \cdot K_r + E_x \cdot (E_x/E_r) \cdot K_x, \quad (1)$$

with K_r, K_x constants depending upon the geometry of the retarding mesh such as shape of the apertures and thickness of the material. For a square mesh of thin material the coefficients are $K_r = 20$ meV and $K_x = 25$ meV with E_r and E_x expressed in V/d units.

Figure 2 shows the calculated transmission function T and its numerical derivative dT/dE for a square mesh aperture as function of the relative kinetic energy of incoming electrons. The relative energy is referenced to the applied retarding potential. The extraction field is set equal to the retarding field $E_x = E_r$. The kinetic energy is measured relative to the mesh voltage. The two particularities of the transmission function are the transmission threshold being very well defined by a sharp rise of T and the noticeable edge effect with T reaching values larger than the expected optical value. The shape of the derivative is nearly triangular with a steep side at the threshold.

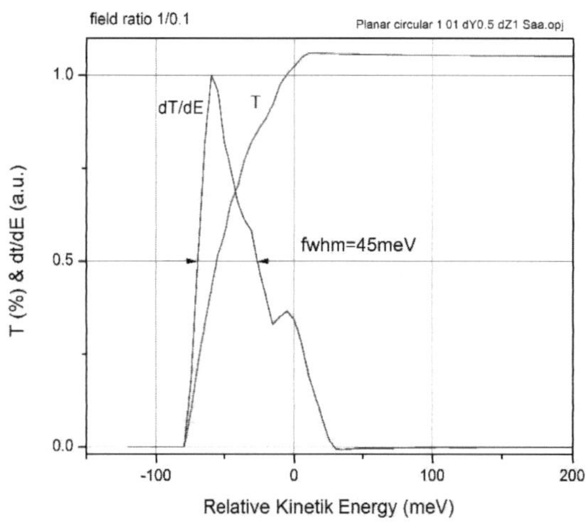

Figure 2. Calculated transmission T and numerical derivative dT/dE.

The expected FWHM of the peak is 45 meV for a field strength of 1 V/d applied on both sides of the mesh. The actual value is obtained scaling with the actual applied field measured in V/d units. The analyser transmission results from the addition of each single square aperture, about 15,000 in this case, and the result will be slightly smoothed according to the geometrical accuracy of the mesh.

5. SEM Imaging with energy filtering of secondary electrons

The analyser is used as secondary electron detector and can be switch between secondary (SE) and backscatter signal (BSE) by adjusting the retarding potential as shown in Figure 3. The PM signal is preferred because of its intensity and achievable scanning speed. The contrast enhancement is maximum when setting the energy to the elastic peak and reveals minute contrast variations not seen in BSE mode when integrating all high energy electrons. Scanning images are also gained using the LI signal at slower scanning speed and have the advantage of clearly separating the single contributions of energy losses as shown for C in Figure 4.

IOP Conf. Series: Materials Science and Engineering **1324** (2025) 012011 doi:10.1088/1757-899X/1324/1/012011

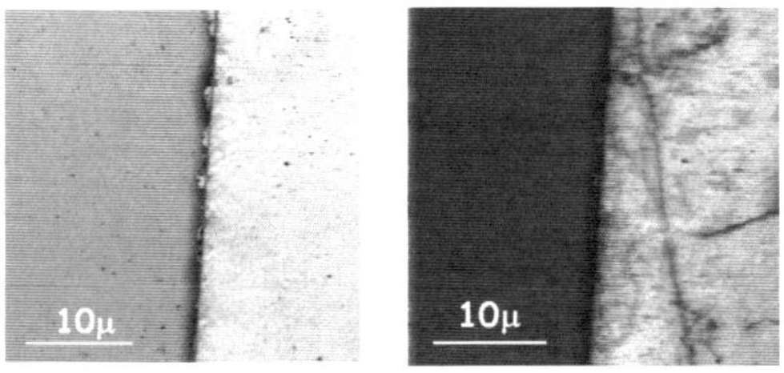

Figure 3. SE at 13 eV and elastic peak at 5,008.0 eV image of Si/Ag. Beam current 7 nA.

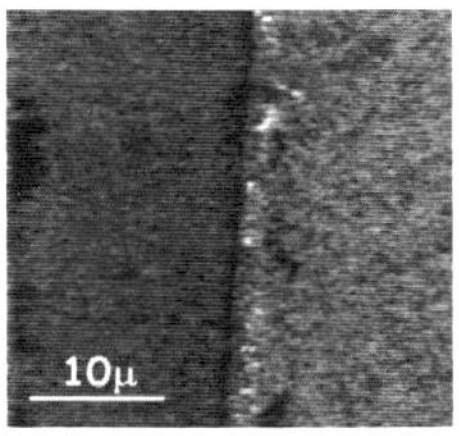

Figure 4. Carbon plasmon distribution at 4,993.0 eV.

6. REELS and EPES applications

The SEM electron gun is a UHV compatible design with a wide primary energy range from 100 eV up to 10 keV. The energy power supply is common to both the SEM and the analyser insuring perfect voltage stability between the two units. The analyser sensitivity allows operation at low primary beam intensity in the nA range or less.

The measured signals are shown in Figure 5 at the elastic peak energy. The direct PM shows a steep increase at the transmission threshold followed by a short plateau and a steady rise at lower analyser energy. The 50 % corresponds to the position of the peak maximum measured by the lock-in and measures the peak intensity. The value at plateau energy measures the integrated signal corresponding to the area under the elastic peak. This is a single measurement instead of an integration done by scanning energy over the peak in the N(E) mode. It can be measured very rapidly, a major advantage for SEM applications.

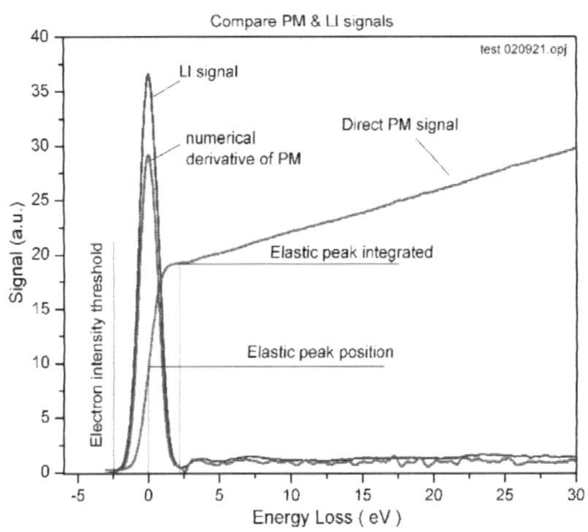

Figure 5. Measured PM signal and its numerical and LI derivatives.

Figure 6 left shows the REELS energy distribution of a clean Ag surface at 5,000 eV and 1.4 nA beam current measured with the lock-in amplifier to differentiate the collected signal. The energy resolution measured at FWHM of the peak is 1.2 eV. The resolution results of several contributions: The intrinsic mesh resolution, the modulation amplitude of the sine voltage applied to the mesh, the spherical aberration of the retarding lens, and the energy spread of primary electron beam. The intrinsic resolution of the analyser is 0.7 eV at 5,000 eV according to the above equation. Figure 6 right uses a logarithmic scale for the intensity with a gauss fit. The dynamic range at the threshold is large and follows the gauss fit over almost 3 decades.

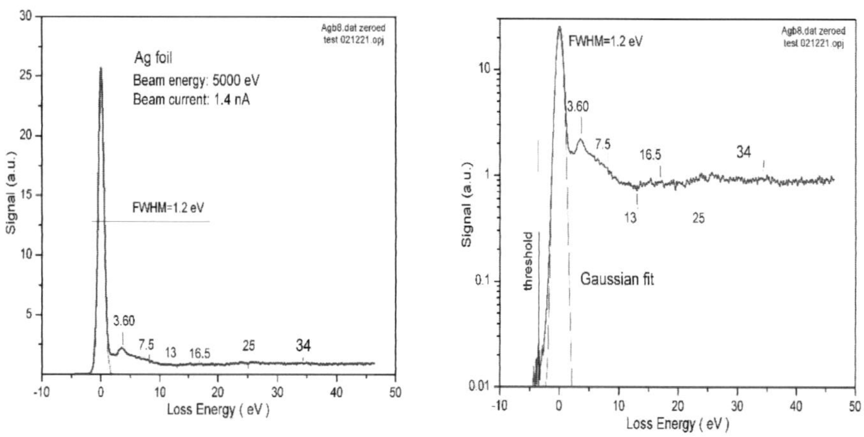

Figure 6. REELS spectra on a clean Ag surface, and right its logarithmic representation.

A further important property of the analyser is the sharpness of the signal drop at the threshold energy. The signal drops several decades when the analyser voltage is raised above the primary beam energy and defines a precisely measurable threshold. Variations of the threshold position are not caused by charging of the surface because the amount of kinetic energy gained or loss by the incoming electrons is compensated by the same loss or gain leaving the surface after backscattering.

7. The M-Z properties of the elastic peak

Shift of the threshold and peak maximum are caused by the recoil energy loss owing to the elastic collision that depends on the atomic mass of the substrate. The process is a Rutherford backscattering event and the broadening of the elastic peak was described in [9]. The recoil energy loss is given by the simplified formula:

$$E_{recoil} = 4(m_e/M)sin^2(\Theta/2)E_p \ , \qquad (2)$$

with M the mass of the recoil atom, m_e the electron mass, Θ the scattering angle and E_p the electron kinetic energy. Recoils loss energies for the present device are plotted in Figure 7. The upper and lower dots indicate the variations due to the wide acceptance angle, and the curve is the mean loss value.

The intensity of the peak is given by the atomic density and the backscattering coefficient can be given by the empirical Everhart formula based on the Rutherford model [10]:

$$\eta = (a - 1 + 0.5^a)/(a + 1) \text{ with } a = \pi Z^2 e^4 N_a / m_e{}^2 dA \ , \qquad (3)$$

where Z is the atomic number, N_a the Avogadro constant, A the atomic weight and d a constant.

Because Z/A is nearly constant for different elements, the backscattering factor essentially reflects the atomic number Z. Thus, the energy shift of the elastic peak measures the atomic mass M of the scattering atom and its intensity correlates to its Z.

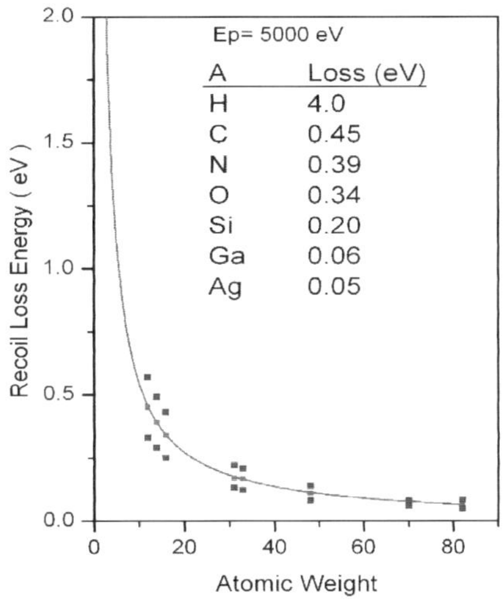

Figure 7. Calculated recoil loss energies at 5 keV primary beam energy and 90° scattering angle.

Figure 8 shows an example of an Au surface contaminated by impurities. The scanning images taken at different energies positions within the elastic peak are different. The SE image is shown on the left for reference, and the images at energies for the peak maximum (marked B) and near the threshold (marked D) reveal several areas of different brightness. The elastic peak shape results from the superposition of different masses of the surface atoms. Imaging the signal at threshold enhance the element of higher mass (Au), whereas lower mass contaminations (C mainly) appear as darker areas.

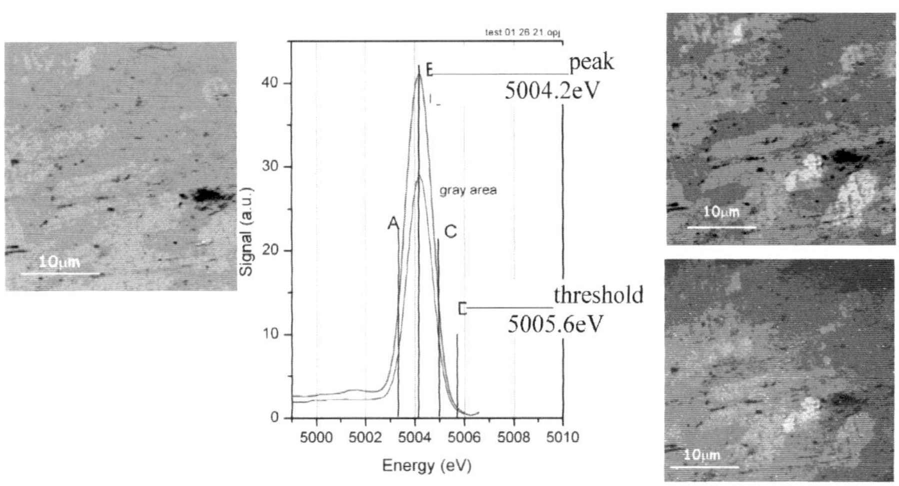

Figure 8. Scanning images of a contaminated Au surface at different energies. Beam current 6 nA.

The analyser can detect low-*Z* materials from H to Si, thus including the main components of biological samples like cellulose from a cell layer of leek (Figure 9). The sheet of cell layer is deposited onto an Ag foil for energy reference and analysed without any further preparation. The left image is from secondary electrons, and the right image of the elastic peak maximum showing the different depth of information. At 5 keV energy the calculated recoil loss values are: ^{108}Ag - 0.05 eV, ^{12}C - 0.45 eV, ^{16}O - 0.34 eV, ^{1}H - 5.4 eV.

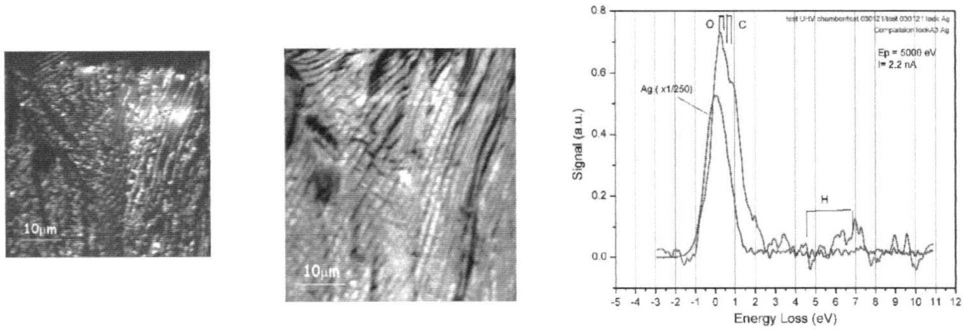

a) b) c)

Figure 9. a) SE image at +13 eV analyser voltage. b) Elastic peak image at 5,002.0 eV. c) Energy losses of cellulose cells and of Ag reference. Beam current 5 nA.

The detection of H is difficult because of the small backscattering cross-section. In addition, the recoil energy deposited to H atoms is large and the atom is likely to be displaced away from its molecular location, thus making the detection of hydrogen essentially destructive when using primary electrons in keV energy range.

8. Detection of photons

Setting the analyser energy above the primary energy suppresses all electrons letting only photons reach the detector. The photomultiplier gain is raised to increase the sensitivity and can be set to a level at which single photon pulses contribute to the signal. Mapping the detector signal reveal areas of cathodoluminescence of insulators and semiconductors. This capability is demonstrated in Figure 10 showing the cathodoluminescent image of phosphorescent powder particles (P14) deposited on a collodion film.

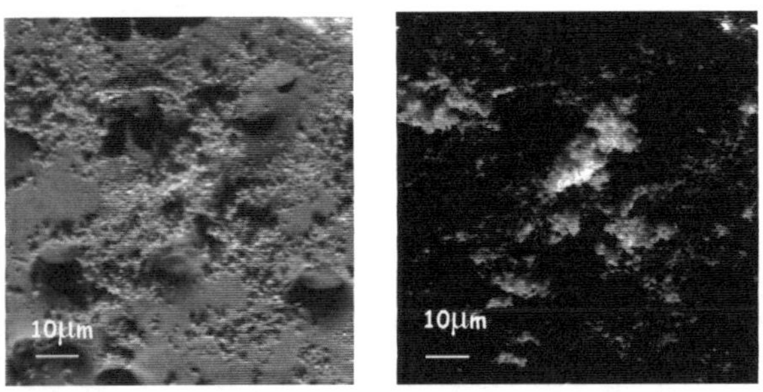

Figure 10. Scanning image at elastic peak energy (4,780 eV) and 14 eV above.

Figure 11 shows a STO sample shortly after treatment with HF acid leaving irregular regions of same composition according to REELS performed at different locations, but slightly different intensities of the elastic peak visible using enhanced contrast to reveal light and dark areas. The intensity variation of the elastic peak is about 2 % and REELS distributions are identical for light and dark regions. The slight change in intensity can be attributed to variations of the inelastic mean free path (IMFP) corresponding to changes of the crystal structure / orientation. The IMFP is characteristic of the material and can be calculated using the NIST data base [8].

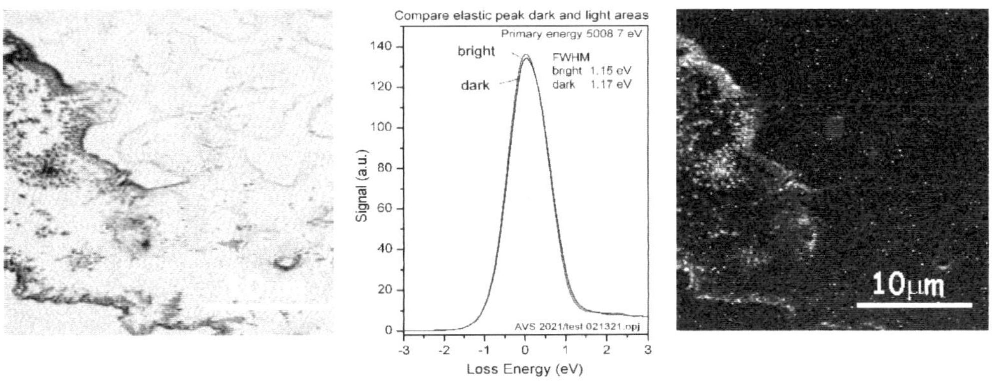

Figure 11. $SrTiO_3$ after etching at elastic peak maximum (left), elastic peaks (centre), and cathodoluminescence image (right).

Soft X-ray emission from GaAs and Si are shown in Figure 12. The metals are mounted above a porous C tape. The photomultiplier works at highest sensitivity in single pulse detection. The beam energy being 5 keV. Carbon can emit an X-ray line from the K-shell ionisation, Si from K- and L-shells, Ga and As from their M-shells. The other contribution is due to the bremsstrahlung continuum integrated of all photon energies, which is low at this energy. The ionisation cross-section for C (K 284 eV) is significant at 5 keV but most recombinations of low energy lead to Auger electron emission rather than soft X-ray. The same yields for all other recombinations such as Si (L_{23} at 99 eV), Si (K at 1,839 eV), Ga (M_1 at 159.5 eV) and As (M_1 at 204.7 eV).

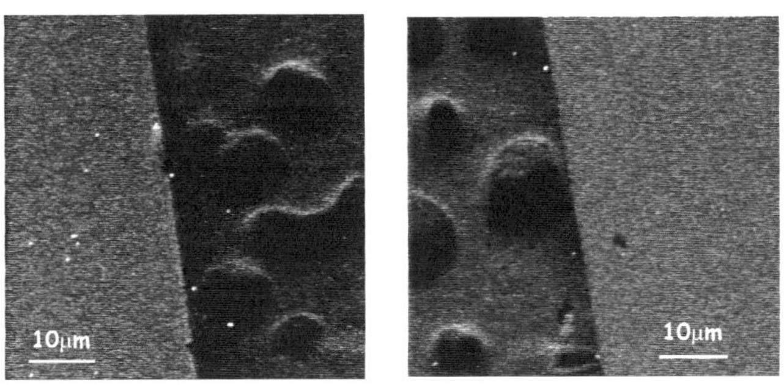

Figure 12. X-ray emission from GaAs on C-tape (left) and Si on C-tape (right) at 6 nA beam current.

9. Operation at higher beam current

Two applications are possible when working at higher beam current in the microampere range: EBSB and AES. EBSD with this design will deliver energy resolved diffraction diagrams but their intensity on the fluorescent screen is low because of the distance between sample and screen, much larger than in usual EBSD setups. Thus the beam intensity must be raised in the μA range unless using a more sensitive detector like channel plates.

The Auger line intensity is much smaller than the CEL signals and the integrated PM signal increases at lower energy making the detection more difficult. AES is however possible using beam currents in the microampere range as the LI signal is shown in Figure 13 for a carbon contaminated silver surface. The energy resolution is excellent and the LMM doublet is fully resolved. The resolution is adjusted by the amplitude of the modulation voltage applied to the retarding electrode.

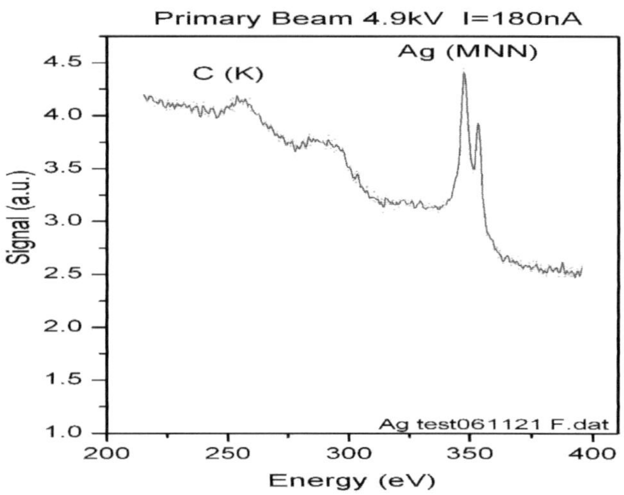

Figure 13. Auger electron spectrum of a carbon contaminated Ag surface. The Ag (MNN) line doublet is fully resolved setting the energy resolution to 1 eV.

10. Conclusions

A new energy analyser designed to fit scanning electron microscopes is described allowing to perform primarily energy filtered secondary and backscattered electron imaging, and also reflection electron energy loss spectroscopy (REELS) and elastic peak electron spectroscopy (EPES). The analyser combines a high sensitivity with high energy resolution and a compact size to fit inside the SEM vacuum chamber. The elastic peak intensity primarily depends on the atomic number of the sample and its energy position of their atomic mass. Lighter elements can be identified by the energy shift of the elastic peak. The analyser performs Auger electron spectroscopy (AES) using higher primary beam currents. Photons are also detected setting the analyser energy above the primary beam energy. Photoluminescence and soft X-ray signals are imaged in scanning mode by filtering out all electrons setting the analyser energy above the primary beam energy.

References

[1] Disko M M, Ahn C C and Fultz B 1992 *Transmission electron energy loss spectroscopy in materials science.* in: EMPMD Monograph Series 2 [Warrendale, PA: Electronic, Magnetic and Photonic Materials Division, Miner. Met. Mater. Soc. (TMS)]

[2] Staib P, Tappe W and Contour J P 1999 *J. Cryst. Growth* **201-202** 45
[3] Staib P J 2011 *Vac. Sci. Technol. B* **29** 3
[4] Deal A and Eades A 2005 *Micros. Microanal.* **11** (Suppl 2) 524
[5] Hooghan T K, Staib P and Eades A 2004 *Micros. Microanal.* **10** (Suppl. 2) 938
[6] Orvis T, Surendran M, Liu Y, Cunniff A and Ravichandran J 2019 *J. Vac. Sci. Technol. A* **37** 061401
[7] Vos M, Marmitt G G, Finkelstein Y and Moreh R 2015 *J. Chem. Phys.* **143** 104203
[8] Powell C J and Jablonski A 2010 *NIST Electron Inelastic-Mean-Free-Path, Standard Reference Database 71*. [Gaithersburg, MD: National Institute of Standards and Technology]
[9] Gergely G, Benedek Zs, Sulyok L, Kover L, Toth J, Varga D, Berenyi Z and Tokesi K 2001 *Vacuum* **61** 107-111
[10] Everhart T E 1960 *J. Appl. Phys.* **31** 1483

Coupled SEM-EDS-RAMAN: A complimentary approach for characterisation – Application to geomaterials

G Wille, D Lahondère and W Kloppmann

B.R.G.M, 3 avenue Claude Guillemin, P.O Box 36009, 45060 Orléans Cedex 2, France

* E-mail: g.wille@brgm.fr

Abstract. A combination of morphological, chemical, and crystallographic analysis is frequently required to characterise materials. Spectroscopic and/or imaging techniques are required to access a thorough knowledge of the sample, to better understand the structural, chemical, and textural factors that control their properties, their behaviour, and their reactivity, ... Among these techniques, scanning electron microscopy and micro-Raman (µ-Raman) spectroscopy occupy a major place and have often been combined. The scanning electron microscope (SEM) provides information on morphology, topography, and chemistry, supplemented by electron beam-based techniques providing chemical or crystallographic information (EDS, WDS, EBSD, cathodoluminescence, etc.). µ-Raman is also widely used in materials characterisation. In geoscience, Raman offers information that includes, among others, polymorphism, crystallography, temperature and stress effects, complementary to that obtained from SEM. Both are often used in parallel. Switching from one to the other is in some cases considered a problem that cannot be solved easily. Then, coupling SEM-EDS and Raman on a unique system is an efficient method for the physicochemical characterisation at the micro/nanometric scale. The technical aspects of coupling are described, as well as practical aspects and limitations inherent in this coupling. The interest will be illustrated by application examples, including cultural heritage, and naturally occurring asbestos (NOA).

1. Introduction

The combination of elemental and crystallographic analysis is frequently required to identify and describe natural and synthetic materials. Physical and chemical characterisation requires a set of analytical spectroscopic and/or imaging techniques to access a thorough knowledge of the sample, to better understand the structural, chemical, and textural factors that control their properties, their behaviour, and their reactivity, etc. As an example, a mineral description often requires chemical information (major, minor, and trace elements), structural information (crystal structure, crystallinity…), and morphological information (crystal shape and size).

Among these characterisation techniques, the scanning electron microscope (SEM) and µ-Raman spectroscopy occupy a major place and have been combined in many studies. The SEM and its various imaging modes (secondary electrons (SE), backscattered electrons (BSE), panchromatic or colour cathodoluminescence (CL), etc.) provide various information on morphology, topography, and chemistry at a nanometre-scale, supplemented by electron beam-based techniques providing chemical or crystallographic information, like energy- or wavelength-dispersive X-ray spectrometry (EDS, WDS), or electron backscatter diffraction (EBSD). µ-Raman spectroscopy is also widely used in organic or inorganic materials characterisation at a micrometre / sub-micrometre scale. The spectrum is unique for a given compound. µ-Raman can provide details on the chemical and structural properties (functional groups, lattice defects, crystallinity, etc.) and various information, which includes, among others, the identification of polymorphs, crystallographic orientation, evaluation of temperature, and stress effects on the structure. Then, both are often used in parallel. Switching from one to the other is in some cases considered a problem (relocation, identification). Coupling SEM-EDS and confocal µ-Raman imaging on the same analytical system is an efficient method for the advanced physicochemical characterisation of materials and the determination of structural properties of materials at the micro-/sub-micrometre scale. The interest of the coupling will be illustrated by various examples of application to geomaterials and mineralogy, including cultural heritage materials, and naturally occurring asbestos, for which this coupling makes it possible to solve problems in the characterisation of such materials.

2. Coupling SEM-EDS and µ-Raman spectroscopy

2.1. Why coupling SEM and µ-Raman spectroscopy

2.1.1. Background and principle of µ-Raman spectroscopy. In 1922, Brillouin [1] and Smekal [2] introduced the theoretical principle of inelastic scattering of light, experimentally demonstrated in 1928 by Raman and Krishnan [3]. However, despite numerous studies on this phenomenon, the so-called Raman effect did not initially succeed in becoming a useful technique for material characterisation. This was due to several technical constraints. During the 1960s, the development of laser sources and the µ-Raman systems, i.e., an optical microscope equipped with a Raman spectroscope allowed new capacities for spectroscopy. With the development of image sensors (charge-coupled devices CCDs), spectrometers and optical assembly, performance, and sensitivity have been improved making Raman spectroscopy a major and powerful analytical technique and an effective tool for analysing natural and synthetic materials. It is widely recognised for acquiring information about molecular vibrations, which provide details of the molecular specificity, structural features, and relatively simple spectral fingerprints of the material. The Raman effect is related to the interactions between light and matter (Figure 1). When photons interact with matter, then 1) part of the light is reflected; 2) part of the light is transmitted; 3) part of the light is absorbed; 4) part of the light is scattered. The diffusion is split into two contributions. A large part of the light keeps its energy, i.e., no energy is exchanged between photons and matter. This is called elastic scattering, or Rayleigh scattering [4] (about 0.1 % of the incident light). About 0.0001 % of the light is scattered with a change in energy, i.e., inelastic scattering also called Raman scattering [3]. The change in wavelength is measured relative to the incident wavelength and termed "Raman shift" (expressed in wavenumbers, cm^{-1}). The Raman shift is associated with two different energy bands. The shift towards wavelengths higher than that of the incident light (lower wavenumber) is termed Stokes scattering. The shift at wavelengths lower than that of the incident light (higher wavenumber) is termed anti-Stokes scattering (Figure 2). Typically, higher-intensity Stokes scattering peaks are used for analysis, but anti-Stokes peaks can also be used.

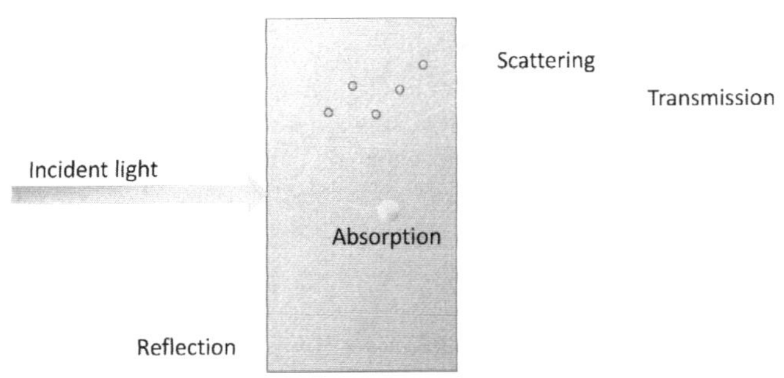

Figure 1. Schematic diagram of light - matter interactions.

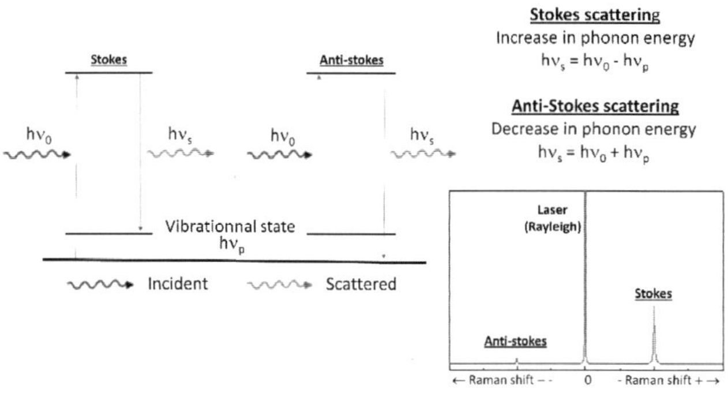

Figure 2. Stokes / anti-Stokes scattering.

The Raman shift is related to the vibrational energy levels in the ground electronic state of the molecule, and as such, the observed Raman shifts are a direct measure of the vibrational energies of the molecule. A Raman spectrum features several peaks, showing the several wavelength positions of the Raman scattered light. Each peak corresponds to a specific molecular bond vibration, including individual bonds and groups of bonds (C-C, aromatic ring, $-CO_3$, –OH, lattice modes, etc.). Raman, therefore, provides a characteristic signature of the analysed phase.

2.1.2. Application of Raman spectroscopy. Raman spectroscopy is used in various fields as a non-destructive, microscopic, chemical analysis and imaging technique. Raman analysis can provide key information to characterise the chemical composition and structure of a sample, whether solid, liquid, or gas. It is for example employed to analyse diverse characteristics and properties of the material: composition, purity [5-7], intrinsic stress/strain [8-10], structure and thickness of 2D materials layers [11-14], identification and characterisation of the typical features of ancient paints and pigments, precious stones, degradation of materials of cultural significance [15-18], etc. In geosciences, Raman applications are countless, for example, identification and distribution of minerals and polymorphs (same chemistry, different crystal structure) [19-21], fluid inclusions in minerals [22, 23], geothermometry [24], etc. Raman spectroscopy is a powerful analytical tool in geology and mineralogy, with advantages over alternative techniques: Little or no sample preparation, microscopic technique complementing conventional petrographic analysis, unambiguous mineral identification, a large variety of sample types (petrographic thin sections, polished samples, powders, piece of rock as it is, etc.). As Raman spectroscopy uses light, it is possible, if a mineral is transparent to the laser wavelength, to analyse under the surface or acquire a stack of spectra/images. However, Raman spectroscopy is subject to some limitations. Not all compounds are Raman active (i.e., produce Raman peaks), or scattering efficiencies can be low for some phases. Raman spectroscopy cannot be used for some materials such as pure metals like gold, silver, and copper, amorphous silica dioxide, diamond, and simple metal salts due to their lack of active Raman vibrations. The Raman effect is very weak (for example, under typical conditions, Rayleigh scattering can be about 10^4 to 10^6 times more intense than Raman scattering), however, these ratio may vary depending on the specific experimental conditions and the nature of the sample) and detection requires sensitive and highly optimised instrumentation. The Raman spectrum can be swamped by fluorescence (due to impurities or the sample itself) from some materials (although this can partly be solved by changing the laser wavelength). The beam energy can destroy or modify the sample. The resolution of an optical microscope in μ-Raman spectroscopy is limited by light diffraction and depth resolution can vary from one sample to another, due to the optical transmittance of each material.

2.1.3. Combining SEM-EDS and μ-Raman for materials characterisation. The interest in combining SEM-EDS and Raman spectroscopy for acquiring interconnected data is established in numerous publications in a large range of applications. A combination of SEM-EDS and μ-Raman is frequently required in material sciences: SEM is a unique tool for imaging samples for morphological, chemical, and structural description, EDS provides the elemental composition, Raman spectroscopy provides structural information on minerals, such as polymorphism, the presence of chemical species (hydroxides, carbonates, organic compounds, fluid inclusions, etc.).

Theiss *et al.* [25] combined SEM-EDS and Raman spectroscopy to identify and elucidate the structure of frapontite, a clay mineral. SEM and EDS are applied to the compositional analysis of the sample, identification of trace elements and purity of the sample (i.e., presence of unexpected phases). Aspects of the molecular structure are obtained by the band assignment of infrared (IR) and Raman spectroscopy. The frapontite "fingerprint" is obtained from the description of the Raman spectrum. Bruni *et al.* [26] combined SEM-EDS and vibrational spectroscopy (Raman and IR) for analysing blue pigments in polychrome artworks. Combining these techniques established the presence of several minerals such as lazurite, nosean, haüyne, and sodalite or their mixtures, which all are constituent parts of the mineral assembly lapis-lazuli. They suggest that their method enables to gain a complete and unambiguous characterisation of the colouring matter. Trace element zoning and incipient metamictisation in a lunar zircon were studied by Wopenka *et al.* [27], by combining secondary ion mass spectrometry (SIMS), electron probe microanalysis (EPMA), and μ-Raman. Major and minor element concentrations were obtained from a thin section of zircon, by EPMA analysis. Trace elements were identified in situ with a SIMS. Raman spectroscopy is applied to study the changes in the crystalline structure of zircon caused by the decay of uranium, and thorium naturally presents in zircons. Raman spectroscopy can also be combined with other SEM-based techniques, like cathodoluminescence (CL) or electron backscatter diffraction (EBSD). Ishihara *et al.* [28] studied the epitaxial growth of graphene on a copper single crystal, to evidence the interest of Cu (1 1 1) for graphene synthesis. For this, the crystallographic orientation of the Cu surface is examined by EBSD before graphene synthesis by chemical vapour deposition (CVD). μ-Raman mapping of the 2D to G peak ratio provides a direct determination of single-layer graphene (SLG) or few-layer graphene (FLG) growth. Zhu *et al.* [29] studied the influence of Nd^{3+}- and Yb^{3+}-dopants on the spectroscopic behaviour of Y_2O_3 ceramic by μ-Raman and CL. Lattice distortion induced by the dopant incorporation and variations of effective absorption coefficients by adding different dopants were assessed using μ-Raman, while the sample homogeneity is evidenced by the local intensity of CL bands. SEM-EDS and μ-Raman spectroscopy were used by Tribaudino *et al.* [30] to explore and understand the mineralogy and texture of an altered Ca-Al-rich inclusion in a carbonaceous chondrite.

The analytical problems presented by the combined use of these two separate devices is evident for these examples. Several authors also noticed the limitations of coupling data from these techniques acquired on separate devices. These can be due to optical microscopy limitations (resolution, depth of field), and difficulties in locating particles of a few tens of μm from one device to the other. These analytical problems may lead to inconsistencies between data from the two techniques. Figure 3 presents the same area observed by light microscopy and SEM to identify observable amphibole particles/fibres. It can be complex to find an area of a few hundred μm on a sample of several cm, with significant differences in visualisation. Moreover, many fibres are visible by light microscopy (LM), whereas they are much less numerous on the backscattered secondary electron (BSE) image. This, therefore, means that many fibres are included in the mineral matrix, and they do not emerge at the surface of the sample. These visualisation differences can lead to misidentification or misinterpretation in the combination of data. The combined use of SEM-EDS and μ-Raman is relevant to characterise these fibres, but the co-location of analyses appears difficult and time-consuming. Several authors have proposed technical solutions to address the problem of particle location. One of the solutions is the use of TEM reference grids or centre-marked grids on a SEM stub as particle support [31-34]. This solution is well suited to micrometric particles, whose size ranges from several micrometres to several tens of micrometres and is in all cases less than the length of the side of the grid holes (54 μm for a 300 mesh grid) and whose composition is unique and homogeneous. However, the problem remains for larger particles (no "large hole" referenced grids), for nanometre particles (too small for LM), or for heterogeneous particles (which is the case for crushed rock for example). Another problem lies in the transport of such a sample from SEM to μ-Raman [35]. A further problematic aspect is related to the depth of field of SEM compared to LM. Thus, particles with strong topography can be easily identified in the SEM but not with LM. Finally, it is important to note that the SEM image is formed from electron-material interactions localised in a thickness of a few nm (SEM) to a few μm (EDS), while the LM can provide details lying several μm, or even several mm below the surface. Thus, errors can be made in the location of the analysed point.

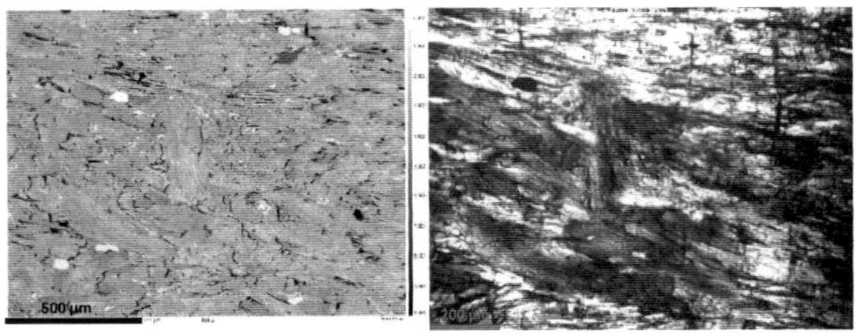

Figure 3. SEM-BSE (left) and light microscopy (right) images of a rock sample. Basic intercalation in glaucophane bearing basic rock (North Corsica, France).

2.2. A short history

Although both techniques have followed a parallel but different evolution since their introduction in the early 1930s, the concept of Raman-in-SEM coupling first emerged in the 1980s. SEM and μ-Raman appear to be essential for the characterisation of the morphology and the elemental and structural chemical properties of materials and minerals. Each technique provides complementary information that can be obtained at the micrometre-scale. The data that each technique provides on its own is often insufficient and has to be combined with data from the other technique. Moreover, compared to optical microscopy, SEM imaging offers a wide range of well-adapted contrasts for distinguishing regions of interest through their chemistry, morphology, crystallography, etc. Their coupling in a unique tool provides a promising analytical combination for such studies. In the 1980s, Truchet and Delhaye proposed and patented a description of an optical system for simultaneous μ-Raman, electron microscopy imaging, and WDS elemental analysis in an electron microprobe (EPMA) or a transmission electron microscope (TEM) [36-38] (Figure 4). They discussed the technical 'state of the art' and the potential interest in biology [39]. EPMA was chosen instead of SEM because of the optical microscope (Cassegrain-type optics) with a micrometre depth of field, placed in the objective lens of the electron optics. They modified the optical system by adding a special device equipped with a semi-transparent plate for sending a laser beam onto the sample surface and transmitting scattered light to the spectrometer. A major obstacle was the very low numerical aperture of Cassegrain objectives, thereby limiting the application field of such μ-Raman spectroscopes to high-scattering samples.

2.3. Integrated solutions for coupling SEM-EDS and μ-Raman microscopy

Following the first developments, the first commercial systems were marketed in the early 2000s. To date, two technical coupling configurations are available on the market [40, 41] (Figure 5).

2.3.1. Co-axial setup. The first technical solution consists of a special interface comprising a parabolic mirror specifically developed for μ-Raman spectroscopy (Figure 6) or a cathodoluminescence mirror (Figure 6d) connected to a spectrometer. The mirror is placed on an automated retractable arm (Figure 6**Figure 5.**a). This system is designed for precisely inserting and retracting the mirror under the

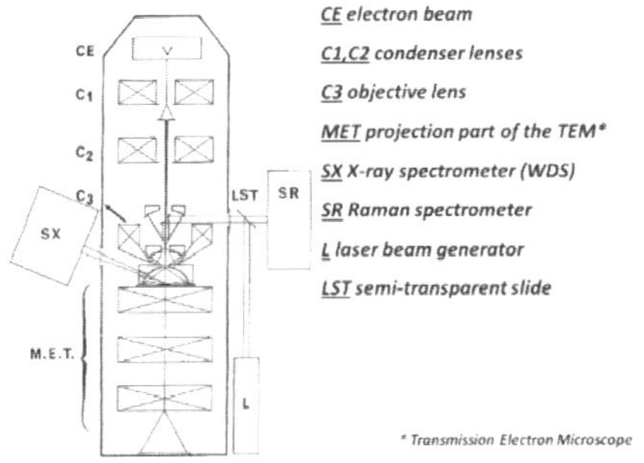

Figure 4. Schematic section of the EPMA column equipped with the μ-Raman microprobe device (after [36]).

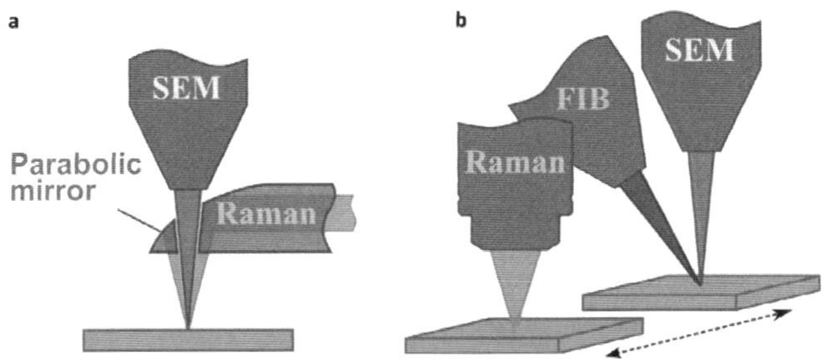

Figure 5. Different SEM - Raman setups. a) Coaxial setup; b) Parallel setup (after [41]).

pole piece of the SEM, between the SEM electron optics and the sample. An aperture in the mirror allows the electron beam to pass through. The laser radiation is focussed on the sample surface by the mirror. In this case, SEM and μ-Raman use are independent. In the first version of this system, the mirror was fixed and allowed only spot acquisition of Raman spectra (Figures 6a and 6b). Recently, new developments have been marketed, with a mobile mirror (3 axes), to acquire 2D and 3D data (Figure 6c). An optical image is collected by the system, making it possible to precisely locate the laser spot on the sample. This image can also be used to check the good agreement between the SEM image and laser beam location. The system enables the collection of a large field-of-view optical image (up to 2 mm × 2 mm) for image correlation. This interface is connected to a conventional μ-Raman spectroscope using optic fibres. The main interest of such a co-axial system is the ability to collect SEM-EDS and Raman data without moving the sample. However, it is necessary to retract the mirror for acquisition like BSE images (the mirror is positioned under the SEM pole piece) or EDS (the EDS detector is masked by the mirror). Some examples of a co-axial setup are shown in Figure 6.

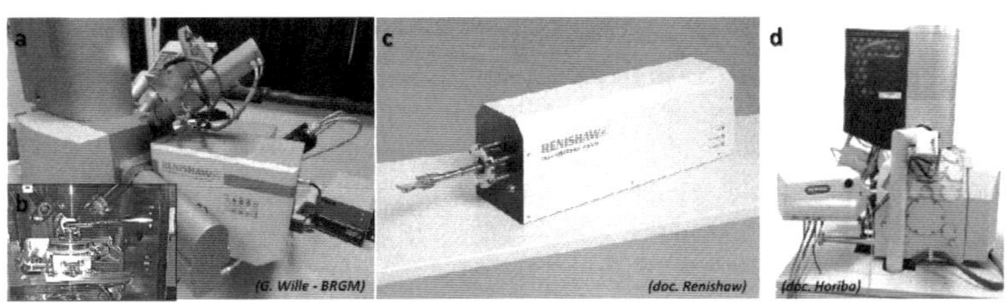

Figure 6. Examples of co-axial setup. a) and b) Raman-in-SEM facility at BRGM (Orleans, France): Renishaw SEM-SCA interfaced on a TESCAN MIRA 3 XMU and mirror in analysis position. c) Renishaw InLux system. d) Horiba RCLUE.

2.3.2. Parallel setup. In the parallel setup, the µ-Raman microscope objective is placed some distance away parallel to the electron column (Figure 7b). The SEM stage is then used for precisely repositioning the sample between Raman and SEM optics. The primary advantage is that the user benefits from the capabilities and performance of a confocal µ-Raman microscope, particularly in terms of resolution, confocality, and signal intensity without affecting the performance of the SEM itself. Some examples of a parallel setup are shown in Figure 7.

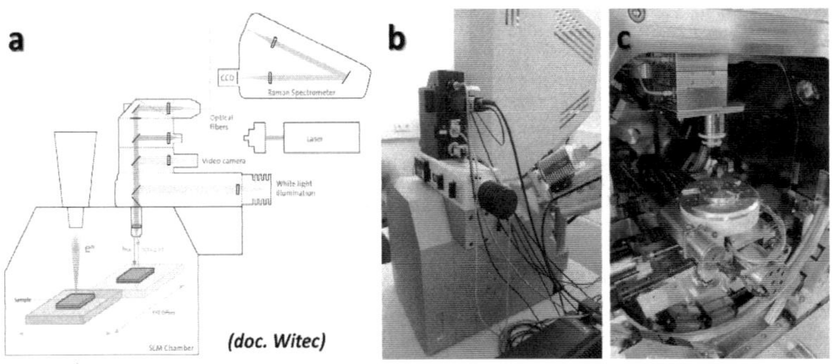

Figure 7. Examples of parallel setup. a) Schematic diagram of the Witec RISE system. b) and c) Witec RISE interfaced on a TESCAN MIRA and interior view of the SEM chamber.

2.4. Which SEM for coupling with Raman spectroscopy

Due to the electron negative charge, SEM samples have to be conductive or coated. To be Raman active, the sample must have vibrations (or rotations) that result in a change in the polarisability of the molecule or material. Since metals are (in first approximation) infinitely polarisable, a vibration of the atoms in the crystal lattice could not cause a change in the polarisability. In the case of metal, Raman diffusion can only be observed for the surface layer (oxidation, contamination), precipitates in alloys (carbides, borides...), ... In the case of non-conductive samples, the accumulation of electronic charges at the point of impact of the beam is known to generate charging effects, which is detrimental to observation and analysis. To avoid having a charge built-up on the surface of non-conductive samples, the surface is usually coated with a conductive layer. When performing EDS analysis, carbon coating is usually recommended. However, carbon coating causes artefacts on the Raman spectrum, which can strongly affect its quality and subsequent data interpretation: a broad band can be observed around 1,500 cm^{-1} and the baseline level is much higher on the coated sample than on the same non-coated sample (Figure 8). The bands are related to the carbon layer: The Raman spectrum of the amorphous carbon layer is characterised by a broad Raman band observed in the range of 1,000 - 1,800 cm^{-1} and this band can be well fitted by two Gaussian distributions known as "D" and "G" bands (band positions: 1,350 and 1,530 cm^{-1} for Scheibe *et al.* [42] or 1,380 and 1,560 cm^{-1} for Marchon *et al.* [43]). The baseline modification is due to a fluorescence phenomenon. The carbon coating behaves as a fluorophore (a compound that causes a molecule to absorb energy at a specific wavelength and then re-remit energy at a different wavelength). Moreover, the presence of the amorphous carbon layer produces a loss of Raman signal intensity and artefacts (fluorescence), as shown in Figure 8 [44]. Under these conditions, the use of a conventional SEM is problematic. Capacities offered by recent FE-SEMs provide good quality images even at low voltage (< a few kV) and make it possible to observe non-coated non-conductive samples in the SEM, but then, EDS microanalysis is difficult or even impossible. Then the use of a low vacuum (or environmental) SEM can be considered the most relevant solution for µ-Raman in SEM.

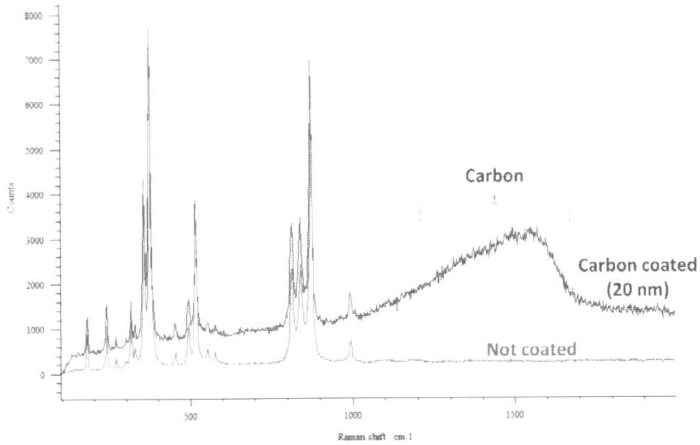

Figure 8. Effect of the carbon coating on the Raman spectrum of an andradite grain (after [44]).

2.5. Cross-influence of the combined use of SEM and μ-Raman

Coupling techniques may cause artefacts directly related to the influence of one on the other. Since Raman-in-SEM spectroscopy can be subject to several influences due to SEM, some possible artefacts, which are due to the impact of the electron beam, can be noticed on the Raman spectra [40, 44].

2.5.1. Contamination. Electron beam-induced contamination has been a well-known problem since the beginning of electron microscopy. It results from the development of carbon deposition on the scanning area. Different sources of contamination have been highlighted and discussed as well as methods to reduce contamination sources and consequences [45]. Since contamination leads to the growth of a thin layer of amorphous carbon on the sample surface, the expected effects are the same as those obtained by conductive carbon layer deposition. An example is shown in Figure 9: A Raman spectrum is collected on cassiterite (SnO_2) after long-time EDS mapping. A broad band is observed at the amorphous carbon bands positions [44]. Its intensity is affected by SEM parameters that influence the development of the contamination layer, like the duration of the electron beam scan, the beam parameters (HV, beam current), etc.

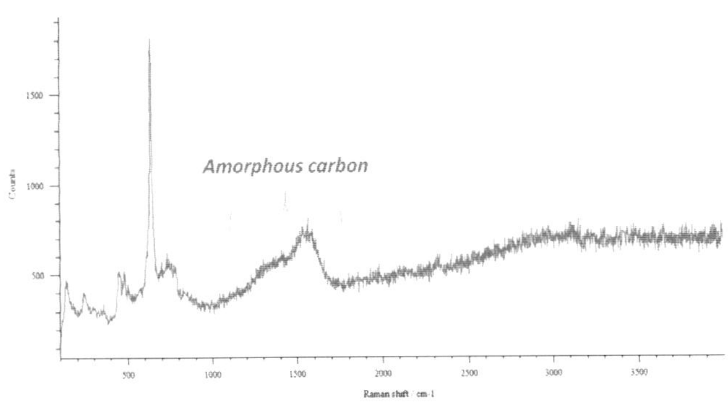

Figure 9. Contamination layer on a cassiterite sample (SnO_2) after a long-time EDS map collection (after [44]).

2.5.2. Cathodoluminescence. Cathodoluminescence (CL) is a photonic emission phenomenon observed when an electron beam (in a SEM, for example) interacts with a sample. CL has many applications in various fields, for example, semiconductors, archaeo-materials, jewellery, mineralogy, and even biological or pharmaceutical samples. CL emission occurs in the visible / near-visible wavelength range, i.e., in the same range as Raman spectroscopy. Therefore, CL can generate artefacts in Raman spectra if not controlled and taken into account. Simultaneous use of both μ-Raman spectroscopy and SEM scanning on a luminescent material causes the appearance of CL bands that are superimposed on the Raman scattering signal [44] as shown in Figure 10.

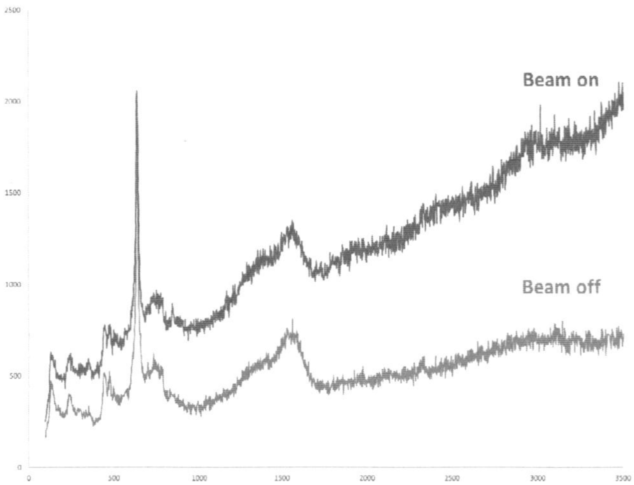

Figure 10. Raman spectrum of cassiterite with (red) and without (blue) electron beam (after [44]).

3. Raman in SEM - Applications

3.1. On-axis setup application on alabaster microstructure of medieval sculptures

Gypsum alabaster is a soft rock, used in medieval and early modern European sculpture, gaining in importance from the 14th to the 17th century, to eventually compete with Italian marble. Chemically, alabaster is made of fine-grained gypsum, a hydrous sulphate of calcium. Its name is derived from another rock used mainly in Antiquity and alternatively named "Egyptian" or "Oriental alabaster", made of calcite, a carbonate of calcium. μ-Raman spectroscopy is one of the techniques that can distinguish gypsum alabaster from calcite alabaster. The purest alabaster is a snow-white material, but it is often associated with other mineral grains (silicates, carbonates, iron oxides), which produce white, grey or coloured clouding and veining in the stone. The softness of alabaster enables it to be carved readily into elaborate forms, but its solubility in water renders it unsuitable for outdoor work. Knowledge of the origin of the alabaster helps to understand the trade routes of this natural material, and isotopic techniques help provide a fingerprint of the different sources of supply [46]. The objective of this study was the description the microstructure and the mineralogy of these veins, to understand their nature and composition and check the possibility of using this information as additional indicator of the origin of the material used for alabaster artwork.

A slice of an alabaster sample from the historical Ickelheim quarries (Bavaria, Germany) [47] was cut and polished. White and coloured veins are visible. Some of these veins were marked so that they can be identified in the SEM (Figure 11a). The veins were investigated by SEM-EDS (Figure 11). The BSE images on the coloured veins highlight different grains of various compositions. The microstructure shows the presence of different minerals that are easily identified by EDS as quartz (SiO_2), iron oxide / hydroxide, as well as calcium carbonates. The map of sulphur shows the presence of grains of calcium sulphate, slightly different from the matrix composed of gypsum. We can assume that these grains correspond to grains of anhydrous calcium sulphate anhydrite ($CaSO_4$).

IOP Conf. Series: Materials Science and Engineering **1324** (2025) 012012 doi:10.1088/1757-899X/1324/1/012012

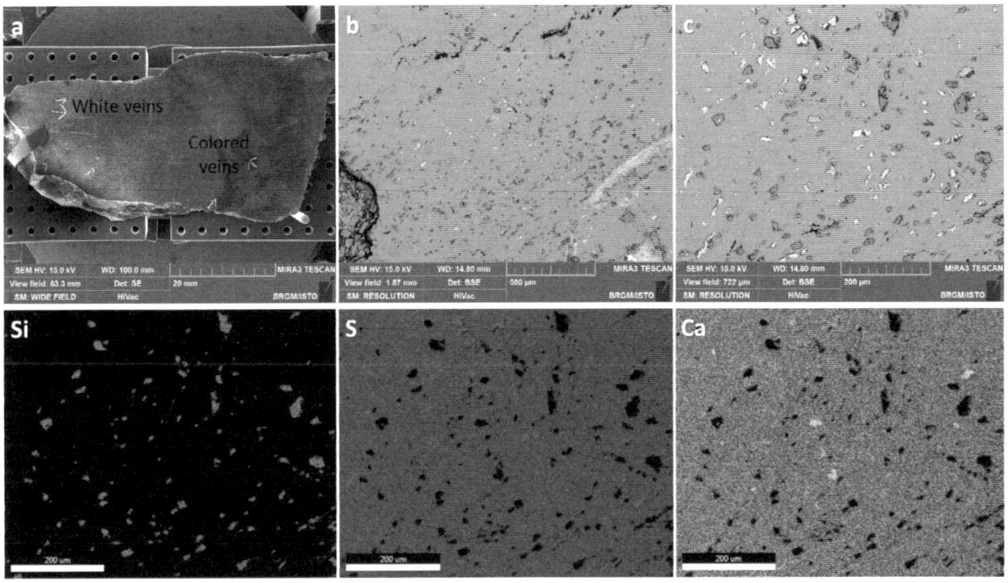

Figure 11. a) SEM (BSE) images of the sample, and b and c) details on a coloured veins. Lower row: EDS mapping (silicon – Si, sulphur – S, calcium – Ca).

These sulphate grains are also present in the white veins. In these veins, we can also note the presence of light grey grains on the BSE image, corresponding to celestine ($SrSO_4$) (Figure 12). The abnormal sulphate grains (slightly brighter than gypsum grains) show similar EDS analyses, and the completion of quantitative analyses is made difficult by the presence of significant porosity of the materials. Further investigations are needed and µ-Raman spectroscopy is a relevant technique for identifying these grains. However, optical microscopy does not make it possible to easily differentiate the different grains. Therefore, the use of a suitable visualisation mode such as BSE in the SEM is required. The Raman spectra of gypsum and abnormal sulphate grains are, therefore, collected in the SEM using a Renishaw SEM-SCA coupled to an InVia spectroscope (Figure **Figure 6.** 6a). Operation conditions were: wavelength 514 nm, power 25 mW, collection duration 200 sec (spectrum accumulation mode).

The comparison of the spectra is presented in Figure 13. The mineralogical identification of unusual phases was performed by searching in a database (RRUFF Database [48] - R040061 record) and confirmed by the bibliographic data [49, 50]. The grains are identified as anhydrite, an anhydrous calcium sulphate. The anhydrous character of the calcium sulphate is highlighted by the absence of Raman bands around 3,500 cm^{-1} (-OH stretching vibration).

3.2. Parallel setup application of high-resolution 2D and 3D imaging on naturally occurring asbestos

Asbestos is a term describing six fibrous silicate minerals that form bundles of long and thin mineral fibres (sub-micrometre diameter). The form and structure of these fibres are called asbestiform. The term asbestos initially refers to minerals marketed and used industrially and, in countries that have regulated or prohibited their use, it is defined by regulation (for example, in Europe, the directive 99/77/EC). The term asbestos usually refers to one fibrous serpentine (chrysotile), and five asbestiform varieties of amphiboles: asbestiform varieties of riebeckite (crocidolite), anthophyllite (anthophyllite-asbestos), grunerite (amosite), actinolite (actinolite asbestos), and tremolite (tremolite-asbestos) [51, 52], even if other minerals can crystallize in a similar habit [53]. If these materials have been (and still are) used industrially, they are initially obtained by extraction in mines,

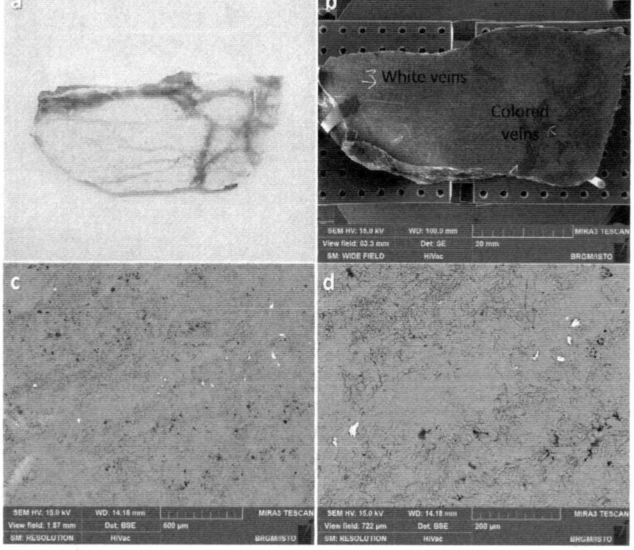

Figure 12. Alabaster sample. a) Photography, and b) SEM-BSE; c and d) Details on a white vein showing celestine grains (brighter grains in the BSE image).

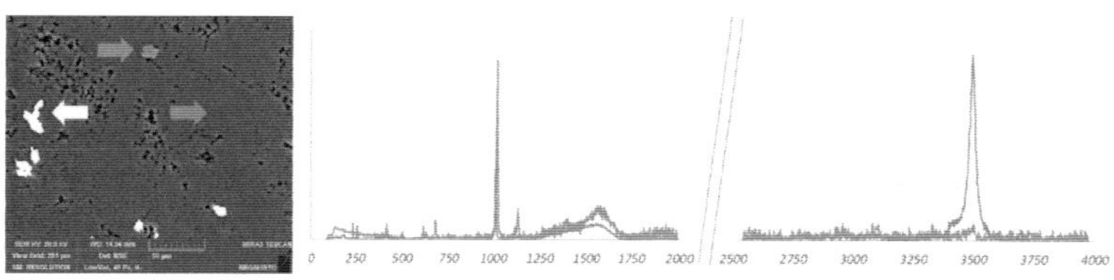

Gypsum	Anhydrite
185	
236	
259	
393	
417	419
494	499
621	612
	629
	679
1010	1019
	1130
3494	

Figure 13. BSE image Raman spectra of gypsum (blue arrow) and the uncommon sulphate grain (orange arrow) - detected Raman bands position (Raman shift, in cm-[1]). Bright grains (white arrow) are celestine $SrSO_4$.

i.e., asbestos initially results from a natural process. Then the term "Naturally Occurring Asbestos" (NOA) refers to fibrous minerals that are found in certain rocks or soil as a result of natural geological processes. Examples of NOA are presented in Figure 14. Natural weathering and routine human activities may disturb NOA-bearing rock or soil and release mineral fibres into the air, which pose a potential risk for human exposure by inhalation. A large set of metamorphic rocks may contain asbestos fibres (meta-ophiolitic rocks of basic or ultrabasic compositions [54-56], hydrothermally altered dolerites [57], talc deposits [58], metamorphosed dolostones and iron formations [59] or sub-alkaline plutonic rocks [60]. Asbestos is considered a Category 1 human carcinogen [52, 54, 55, 61, 62], as inhalation of asbestos fibres causes respiratory diseases, in particular asbestosis, lung cancers, and malignant mesothelioma.

Optical microscopy is the most currently used method for the regular measuring of asbestos, although it does not allow identification and counting of individual fibres detrimental to health [63]. Transmission electron microscopy (TEM) is the most efficient method for asbestos diagnosis in air, due to its capacity to provide morphology, chemistry, and crystallographic information together on the same fibre [64]. But in the case of amphibole included in a mineral matrix, TEM is hard to apply: It is likely to release artefacts like cleavage fragments due to sample crushing, and sample-piece thinning by ion beams

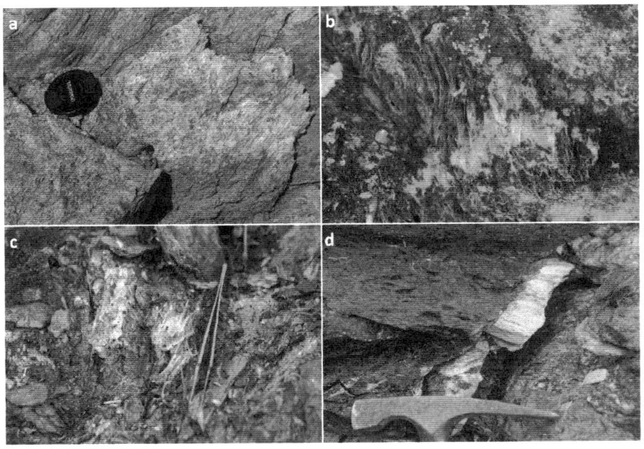

Figure 14. Examples of naturally occurring asbestos. a) "slip-vein" with chrysotile cutting serpentinites. b) Polymineral vein with actinolite-asbestos cutting ophiolitic metagabbro. c) Clusters of tremolite-asbestos fibres in mylonitic. d) Cataclastic serpentine (photographs: G. Wille, BRGM).

techniques is costly and hard to achieve and thus not applicable in routine analysis. In addition, the differentiation of real asbestos fibres and cleavage fragments of non-asbestos amphiboles (created by a grinding step in the protocol) is complex in many cases and the criteria proposed by different authors, if they are often relevant, are not completely discriminating [65-67] and can lead to "false positive" result. In addition, it can also be questioned about the representativeness of the sampling of some cubic micrometres compared to the heterogeneity of a sample at the scale of several millimetres or even centimetres. Then the development of an alternative analysis protocol, adapted in terms of spatial resolution and mineralogical discrimination capabilities, and which does not generate sample preparation artefacts, is relevant.

Among the methods that meet these criteria (nature and morphology of the mineral fibres), the combined use of SEM-EDS and µ-Raman spectroscopy is a relevant solution [68, 69]. Indeed, the resolution of SEM imaging makes it possible to easily observe fibres of less than a few hundred nm in diameter. Combined with EDS analysis, µ-Raman spectroscopy can provide a mineralogical signature of the serpentine [70-72] or amphibole [73, 74]. But, above all, these techniques, combined within the same analytical system, can be applied to a coherent rock sample (polished section or thin geological section), without prior grinding (therefore, without generating cleavage fragments linked to the sample preparation [66, 75-78]). Confocal µ-Raman-in-SEM can be performed on any polished SEM sample. Samples were prepared as thin geological slices: A block of rock is glued on a glass slice then thinned and polished to a thickness of 40 µm. The samples were observed by polarised light microscopy (PLM) before SEM analysis, for a petrographic study and location of regions of interest (i.e., presence of fibrous / non-fibrous amphiboles). The identification of amphiboles was confirmed by EPMA analyses on the micrometre particles/bundles. Coupled SEM observations and Raman imaging in SEM were performed on a TESCAN-Witec RISE microscope combining a TESCAN Mira 3 GMU LV-FE-SEM coupled to a WITec confocal µ-Raman imaging system. The RISE confocal µ-Raman imaging system is equipped with a UHTS300 spectrometer, a Zeiss 100× vacuum objective (numerical aperture N.A. = 0.75) mounted inside the SEM chamber, using a 532 nm laser radiation wavelength. Coupled SEM-EDS and Raman imaging was performed on non-coated samples at HV = 15 kV under low vacuum conditions (nitrogen, pressure 20 - 40 Pa).

An example is presented in Figure 15. A thin geological slice was prepared on a rock from a stone quarry. EPMA analyses prove the presence of actinolite but does not make it possible to precisely identify the thinnest particles with a diameter of less than 1 µm. Confocal µ-Raman is collected in the

SEM: The thinnest fibres are located using the SEM imaging and EDS analysis. (Figures 15a and 15b), then a Raman map is collected on the same area light microscopy image is first collected (Figure 15c) the Raman mapping is performed on a selected area. After data analysis, the spectra of the two phases present are identified by analysis of the data cube, allowing the production of a phase map and the corresponding spectra (Figures 15d and 15e). The confocal Raman mapping makes it possible to identify actinolite fibres in an orthoclase matrix, including the thinnest fibres, with a diameter of 400 nm or less.

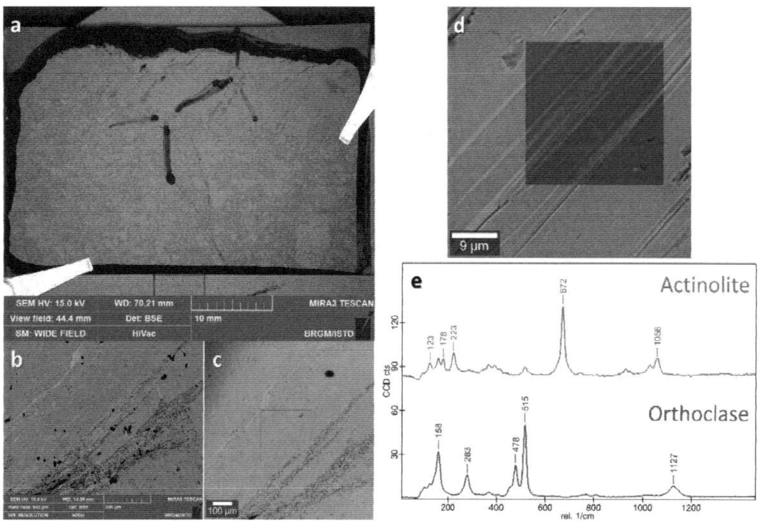

Figure 15. Confocal Raman-in-SEM analysis of actinolite fibres in an orthoclase matrix (south of the Armorican Massif, France).

A confocal microscope is a system for which illumination and detection are limited to the same volume of reduced size. The resolution is directly related to the value of the numerical aperture (N.A.). When the system is highly confocal, it allows a short resolution in Z. If the material is transparent to the laser radiation, the confocal Raman-in-SEM also allows the acquisition of 3D measurements by combining a series of maps along the Z-axis. An example is presented in Figure 16. A series of maps are collected with a step of 0.5 µm from the surface to 20 µm then combined to produce a 3D volume. In Figure 16a, a 2D map is presented. Actinolite particles or fibres are visible on the surface. The interest of the 3D view is to provide information on the true shape of the particles of actinolite in the matrix and to make it possible to determine a volumetric quantity of fibres, a data very difficult to establish due to the nanometre dimension of the fibres. For example, on the 3D map, a short elongated particle of actinolite on the 2D map (circle) is detected. Using the 3D view, it is possible to identify this particle as being in fact, a fibre positioned obliquely to the surface of the sample.

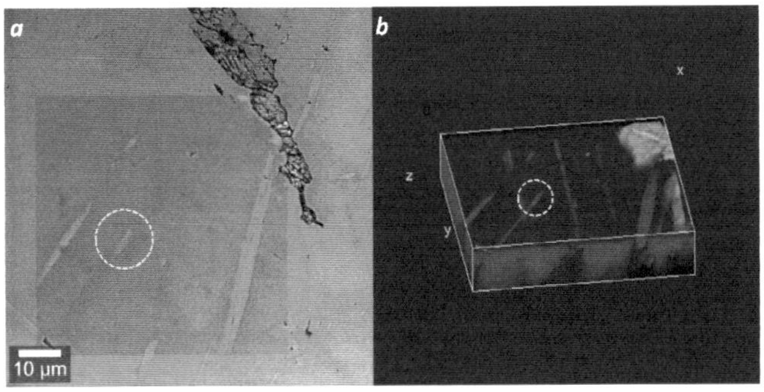

Figure 16. 3D Raman-in-SEM: actinolite fibres (red) in albite (blue) (after [68]).

4. Conclusion

The interest in combining SEM-EDS and µ-Raman spectroscopy for the characterisation of materials is highlighted by numerous authors for many years. The emergence of devices that integrate these two techniques in a single analytical system has revealed new capabilities for co-located data correlation. Raman-in-SEM is a powerful combination of two mature technologies, which results in a set of chemical and structural data that redefine convenience, efficiency, and productivity in the characterisation of materials. Numerous applications are mentioned in the literature, including materials sciences, industrial processing, nanotechnologies, corrosion or oxidation of reactive phases, multilayer deposits, life sciences, etc. In geosciences, the combination of elemental and crystallographic analysis is frequently required to identify and describe a mineral. Mineralogical analysis requires a set of analytical imaging techniques to access a thorough knowledge of the sample and better understand the mineralogical, chemical, and textural factors that control their behaviour and reactivity. Among these techniques, SEM and µ-Raman spectroscopy occupy a major place among the characterisation techniques. SEM provides various information on morphology, topography, and chemistry at a nanometre scale, supplemented by elemental and crystallographic analysis. Raman spectrum is a sort of fingerprint of a given compound and the technique is sensitive to various elements of the local environment. It can reveal a mineral's crystal structure, providing details on its chemical and structural properties. µ-Raman offers various information that includes, among others, the identification of polymorphs, crystallographic orientation, evaluation of temperature, and stress effects on the structure. Moreover, µ-Raman provides complementary chemical and structural information to that obtained from SEM. This is why both are often used in parallel. Then, coupling SEM-EDS and µ-Raman spectroscopy and imaging on the same analytical system appears to be an efficient method of identifying the mineral nature and determining the structural properties of geomaterials. This multi-technique microscopy allows the collection of a real mineralogical analysis of rocks. This coupling makes it possible to solve problems for the mineralogical characterisation of geomaterials.

Acknowledgments
The authors would like to thank their colleagues and ex-colleagues who participated in this work, Ms J. Duron, Dr C. Lerouge, Dr X. Bourrat. We thank peoples from Witec and Tescan Dr U. Schmidt, Dr P. Ayasse, Dr J. Silvent, Mr D. Barresi who allowed us to perform RISE and SEM-FIB analyses on the NOAs at Witec (Ulm, Germany) and Tescan France (Fuveau, France) application labs. Dr M. Tchaya (WITec), Mr J. Brites (Horiba) and Mr M. Belleil (Renishaw) are thanked for their help and technical discussions, and for providing interesting documents and application on the coupling systems and their evolutions. The study on alabaster was co-funded by the Louvre Museum and BRGM.

References

[1] Brillouin L 1922 *Ann. Phys. (Paris)* **9** 88-122
[2] Smekal A 1923 *Naturwissenschaften* **11** 873-875
[3] Raman C V and Krishnan K S 1928 *Nature* **121** 501-502
[4] Rayleigh, Lord 1881 *London, Edinburgh, Dublin Philos. Mag. J. Sci.* **12** 81-101
[5] Begliarbekov M, Sul O, Kalliakos S, Yang E-H and Strauf S 2010 *Appl. Phys. Lett.* **97** 031908
[6] Fischer S, Sturm M, Schlösser M, Bornschein B, Drexlin G, Priester F, Lewis R J and Telle H H 2011 *Fusion Sci. Technol.* **60** 925-930
[7] Jamur J 2022 *Spectrosc. Eur.* **34** 18-22
[8] Angel R J, Murri M, Mihailova B and Alvaro M 2019 *Z. Krist. - Cryst. Mater.* **234** 129-140
[9] Ma L, Qiu W and Fan X 2021 *Microelectron. Reliab.* **118** 114045
[10] Liu D and Flewitt P E J 2014 Raman measurements of stress in films and coatings. in: *Spectroscopic properties of inorganic and organometallic compounds: Techniques, materials and applications. Vol. 45.* (Douthwaite R, Duckett S and Yarwood J; Eds.) [London, UK: Royal Society of Chemistry] 141-177
[11] Tan P-H 2019 *Raman spectroscopy of two-dimensional materials.* in: Springer Series in Materials Science, vol. 274. [Singapore: Springer Singapore]
[12] Cong X, Liu X-L, Lin M-L and Tan P-H 2020 *Npj 2D Mater. Appl.* **4** 13
[13] Taghizadeh A, Leffers U, Pedersen T G and Thygesen K S 2020 *Nat. Commun.* **11** 3011
[14] Malard L M, Lafeta L, Cunha R S, Nadas R, Gadelha A, Cançado L G and Jorio A 2021 *Phys. Chem. Chem. Phys.* **23** 23428-23444
[15] Casadio F, Daher C and Bellot-Gurlet L 2016 *Top. Curr. Chem.* **374** 62
[16] Lofrumento C, Zoppi A and Castellucci E M 2004 *J. Raman Spectrosc.* **35** 650-655
[17] Gueriau P, *et al.* 2017 *Anal. Chem.* **89** 10819-10826
[18] Colomban P 2004 *Appl. Phys. A* **79** 167-170
[19] Armbruster T, *et al.* 2006 *Eur. J. Mineral.* **18** 551-567
[20] Hope G A, Woods R and Munce C G 2001 *Miner. Eng.* **14** 1565-1577
[21] Griffith W P 1969 *Nature* **224** 264-266
[22] Frezzotti M L, Tecce F and Casagli A 2012 *J. Geochemical Explor.* **112** 1-20
[23] Caumon M, Tarantola A and Wang W 2020 *J. Raman Spectrosc.* **51** 1868-1873
[24] Beyssac O, Goffé B, Chopin C and Rouzaud J N 2002 *J. Metamorph. Geol.* **20** 859-871
[25] Theiss F L, López A, Scholz R and Frost R L 2015 *Spectrochim. Acta Part A Mol. Biomol. Spectrosc.* 147 230-234
[26] Bruni S, Cariati F, Casadio F and Toniolo L 1999 *Vibr. Spectrosc.* **20** 15-25
[27] Wopenka B, Jolliff B L, Zinner E and Kremser D T 1996 *Am. Mineralogist* **81** 902-912
[28] Ishihara M, Koga Y, Kim J, Tsugawa K and Hasegawa M 2011 *Mater. Lett.* **65** 2864-2867
[29] Zhu W, Wu Y, Leto A, Du J and Pezzotti G 2013 *J. Phys. Chem. A* **117** 3599-3607
[30] Tribaudino M, Bersani D, Mantovani L, Pizzati M and Salviati G 2021 *J. Raman Spectrosc.* **52** 1892-1901
[31] Stefaniak E A, Worobiec A, Potgieter-Vermaak S, Alsecz A, Török S and Van Grieken R 2006 *Spectrochim. Acta Part B Atom. Spectrosc.* **61** 824-830
[32] Stefaniak E A, Alsecz A, Frost R, Máthé Z, Sajó I E, Török S, Worobiec A and Van Grieken R 2009 *J. Hazard. Mater.* **168** 416-423
[33] Pointurier F and Marie O 2010 *Spectrochim. Acta Part B Atom. Spectrosc.* **65** 797-804.
[34] Godoi R H M, Potgieter-Vermaak S, De Hoog J, Kaegi R and Van Grieken R 2006 *Spectrochim. Acta Part B Atom. Spectrosc.* **61** 375-388
[35] Nelson M P, Zugates C T, Treado P J, Casuccio G S, Exline D L and Schlaegle S F 2001 *Aerosol Sci. Technol.* **34** 108-117
[36] Truchet M and Delhaye M 1986 *Dispositif de microscopie analytique propre à former à la fois une sonde Raman et une sonde électronique.* [Paris, France: C.N.R.S. Inpi (Ed.)]

[37] Truchet M and Delhaye M 1988 *J. Microsc. Spectrosc. Electron.* **13** 167-197.
[38] Dhamelincourt P, Delhaye M, Truchet M and da Silva E 1991 *J. Raman Spectrosc.* **22** 61-64
[39] Truchet M, Delhaye M, Ballan-Dufrançais C and Jeantet A-Y 1988 *Biol. Cell* **63** 26
[40] Wille G, Bourrat X, Maubec N, Guegan R and Lahfid A 2014 Raman-in-SEM studies of inorganic materials. in: *Spectroscopic properties of inorganic and organometallic compounds: Techniques, materials and applications. Vol. 45.* (Douthwaite R, Duckett S and Yarwood J; Eds.) [London, UK: Royal Society of Chemistry] 79-116
[41] Jiruše J, Haničinec M, Havelka M, Hollricher O, Ibach W and Spizig P 2014 *J. Vac. Sci. Technol. B, Nanotechnol. Microelectron. Mater. Process. Meas. Phenom.* **32** 06FC03
[42] Scheibe H-J, Drescher D and Alers P 1995 *Fresenius. J. Anal. Chem.* **353** 695-697
[43] Marchon B, Jing Gui, Grannen K, Rauch G C, Ager J W, Silva S R P and Robertson J 1997 *IEEE Trans. Magn.* **33** 3148-3150
[44] Wille G, Bourrat X, Maubec N and Lahfid A 2014 *Micron* **67** 50-64
[45] Postek M T 2006 *Scanning* **18** 269-274
[46] Kloppmann W, *et al.* 2017 *Proc. Natl. Acad. Sci.* **114** 11856-11860
[47] Kloppmann W, Leroux L, Bromblet P, Le Pogam P-Y, Montech A T and Guerrot C 2022 *PLoS One* **17** e0265242
[48] Lafuente B, Downs R T, Yang H and Stone N 2016 The power of databases: The RRUFF project. in: *Highlights mineralogy and crystallography;* (Armbruster T and Danisi R M; Eds.) [De Gruyter] 1-30
[49] Berenblut B J, Dawson P and Wilkinson G R 1971 *Spectrochim. Acta Part A Mol. Spectrosc.* **27** 1849-1863
[50] Berenblut B J, Dawson P and Wilkinson G R 1973 *Spectrochim. Acta Part A Mol. Spectrosc.* **29** 29-36
[51] WHO 1986 *Asbestos and other natural mineral fibres, environmental health criteria.* [Geneva, Switzerland: WHO]
[52] National Institute for Occupational Safety and Health (NIOSH) 2011 Curr. Intell. Bull. **153**
[53] Lockey J 1981 *Clin. Chest Med.* **2** 203-218
[54] Vignaroli G, Ballirano P, Belardi G and Rossetti F 2014 *Environ. Earth Sci.* **72** 3679-3698
[55] Campopiano A, *et al.* 2018 *Int. J. Environ. Health Res.* **28** 134-146
[56] Lahondère D, Cagnard F, Wille G and Duron J 2019 *Environ. Earth Sci.* **78** 540
[57] Lahondère D, Cagnard F, Wille G, Duron J and Misseri M 2018 *Environ. Earth Sci.* **77** 385
[58] Van Gosen B S, Lowers H A, Sutley S J and Gent CA 2004 *Environ. Geol.* **45** 920-939
[59] Van Gosen B S 2007 *Environ. Eng. Geosci.* **13** 55-68
[60] Metcalf R V and Buck B J 2015 *Geology* **43** 63-66
[61] U.S. Department of Labor, Mine Safety and Health Administration (MSHA) 2005 *Preliminary regulatory economic analysis for proposed rule on asbestos exposure limit.* [Washington, DC: U.S. Department of Labor, Mine Safety and Health Administration (MSHA), Office of Standards, Regulations, and Variances] RIN: 1219-AB24 https://arlweb.msha.gov/REGS/REA/05-14510(Asbestos).pdf
[62] Lahondère D and Cagnard F 2018 in: *XIII IAEG Congress - Engineering Geology for a Sustainable World.* (San Francisco, CA: September 2018)
[63] Boulanger G, *et al.* 2014 *Environ. Health* **13** 59
[64] Steen D, Guillemin M P, Buffat P and Gilbert L 1983 *Atmos. Environ.* **17** 2285-2297
[65] Van Orden D R, Allison K A and Lee R J 2008 *Indoor Built Environ.* **17** 58-68
[66] Belardi G, Vignaroli G, Trapasso F, Pacella A and Passeri D 2018 **10** 63-78
[67] Harper M, Lee E G, Doorn S S and Hammond O 2008 *J. Occup. Environ. Hyg.* **5** 761-770
[68] Wille G, Lerouge C and Schmidt U 2018 *J. Microscopy* **270** 309-317
[69] Wille G, Lahondère D, Schmidt U, Duron J and Bourrat X 2019 *J. Hazard. Mater.* **374** 447-458
[70] Groppo C, Rinaudo C, Cairo S, Gastaldi D and Compagnoni R 2006 *Eur. J. Mineral.* **18** 319-329

[71] Rinaudo C, Gastaldi D and Belluso E 2003 *Can. Mineralogist* **41** 883-890
[72] Petriglieri J R, Salvioli-Mariani E, Mantovani L, Tribaudino M, Lottici P P, Laporte-Magoni C and Bersani D 2015 *J. Raman Spectrosc.* **46** 953-958
[73] Leissner L, Schlüter J, Horn I and Mihailova B 2015 *Am. Mineralogist* **100** 2682-2694
[74] Bersani D, Andò S, Scrocco L, Gentile P, Salvioli-Mariani E, Fornasini L and Lottici P P 2019 *Minerals* **9** 491
[75] Langer A M and Nolan R P 1987 *N. Engl. J. Med.* **316** 882-882
[76] Bloise A, Kusiorowski R and Gualtieri A 2018 *Minerals* **8** 274
[77] Baietto O, Diano M, Zanetti G and Marini P 2019 *Fibers* **7** 52
[78] Militello G M, Sanguineti E, Gonzàlez A and Gaggero L 2020 *Episodes* **43** 909-918

Focussed beams for use in EBSD

B Winiarski

Thermo Fisher Scientific, Materials & Structural Analysis, Pecha 12, 62700 Brno, Czech Republic

E-mail: bartlomiej.winiarski@thermofisher.com

Abstract. Focussed beams (electrons, ions, and photons) are at the forefront of today's and future scientific and engineering advanced research instruments. The beams are typically used for six purposes: imaging, detection, deposition, implantation, milling/ablation, and polishing. There are plethora of detectable and useful signals coming from the beam irradiated matter. One of the signals is electron diffraction and scattering on a crystal planes used by electron backscatter diffraction (EBSD) method. This surface superficial probing technique is sensitive to surface contamination, amorphisation, and lattice distortions, thus requires damage-free sample surface preparation techniques. The advent of coupled multi-beam commercial instruments (DualBeams, CrossBeams and TriBeams) allowed for integration of sample surface preparation techniques, i.e., site specific (cross-sections, serial sections, thin lamellas, micro- and nano-pillars, etc.), and large areas (cross-sections and surface polishing), and beam excited signal detection/quantification methods, i.e., EBSD, EDS, WDS, SIMS, Raman, LIBS, etc. This paper attempts to bring closer the current state-of-the-art of focussed beams systems for use in EBSD technique in Materials Science.

1. Historical brief on development of focussed beams

Focussed beams are complex and compact systems that emit, accelerate, deflect, and raster a narrow stream of: *i*) negatively charged particles - electrons, *ii*) positively charged particles - ions, or *iii*) photons. Currently these beams are typically used for six purposes: imaging, detection, deposition, implantation, milling/ablation, and polishing.

In the early days these devices formed a standalone analogue transmission electron microscope (TEM) [1, 2], scanning electron microscope (SEM) [3, 4], scanning ion microscope (SIM) (commercial systems based on liquid Ga^+ metal ion source (LMIS)) [5, 6], and light amplification by stimulated emission of radiation (LASER) (this is a very broad topic; here it is referred to low power < 10 W and pulsed lasers with impulse duration < 1 ns time domains, having application for low thermal or near-athermal material micro machining, laser-induced signal measurements, e.g., laser-induced breakdown spectroscopy, and laser assisted material deposition) [7, 8], or short pulse (SPL) optics systems [9]. Currently, these standalone microscopes are based on digital circuits/logic systems. TEMs and SEMs evolved with improved electron sources, electron optics, aberration correctors, and various detection systems [10, 11]. SIM platforms range was extended with commercial scanning helium ion microscopes (SHIM) in 2008 [12] (also, neon is possible as primary ions), and xenon plasma focussed ions beam (PFIB) in 2011 [13]. Several experimental FIB sources are developed as well [14-16], e.g., liquid metal alloy ion sources (LMAIS), gas field ion sources (GFIS), low temperature ion sources (LOTIS), and magneto optical trap ion sources (MOTIS).

From the mid-2000's pulsed lasers and optics for fast micro-machining at sub mm-scale and in situ laser-based characterisation techniques have moved to ultra-short pulse lasers (UPL) (having a pulse duration < 1 ps, typically ~ 50 - 300 fs). These laser systems have capabilities of athermal and near damage-free ablation with the laser spot size of dozens of micrometres [18].

Circa the early 1990's the first systems with two columns emerged commercially [19]. These platforms have merged capabilities of SEM and focussed ion beam (FIB) column (based on liquid Ga^{+} metal ions source – LMIS [5]), forming a single microscope with two beams defining a coincidence point (Figure 1). In so doing a high-resolution electron imaging and qualitative detection modes of SEM (secondary electrons and backscattered electrons detection) are combined with implantation and milling/polishing mode of FIB. This configuration allows for site-specific micrometre-scale work, e.g., cross-sectioning, serial sectioning tomography, etc. [17]. Gradually, FIB-SEMs have been upgraded with analytical capabilities, e.g., energy dispersive X-ray spectrometry (EDS) and electron backscatter diffraction (EBSD). Automations of FIB-SEM routines led to development of 3D EBSD and 3D EDS methods (Figure 2).

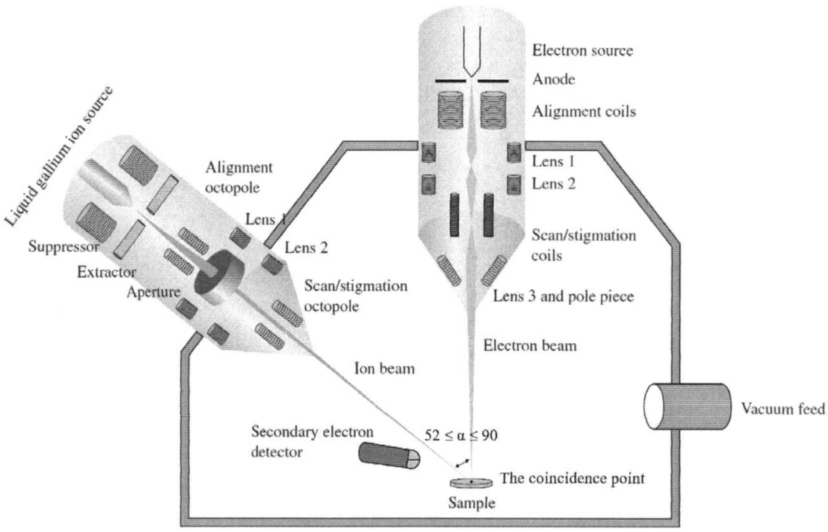

Figure 1. A schematic diagram is showing the configuration of two-beam focussed ion beam – scanning electron microscope system [17].

FIB at low beam currents (pA) can be focussed to a few nm spots sizes. However, the use of Ga^{+} FIB column has two drawbacks:

a) Gallium can create low temperature eutectic alloys during ion matter interaction, particularly present along grain boundaries [21-23], e.g., in Figure 3. Therefore, some Al, Cu, Fe and Ni alloys are prone to liquid metal embrittlement [24] during FIB irradiation.
b) LMIS is divergent ion source (Figure 4a) emitting a beam shaped and focus to a spot by a FIB column. This limits the optimal beam shape and the maximum current density that is achieved by a FIB column. Therefore, usable beam shapes, beam spot sizes, and beam current densities are limited to beams having currents < 65 nA [25], see Figure 4c. These constrain the maximum material removal rate (e.g., 17.6 $\mu m^3/s$ for Si [26]) and the maximum volume of material (for most of the materials $< 50 \times 50 \times 50\ \mu m^3$, e.g., Figure 2c) that is removed with reasonable amount of time.

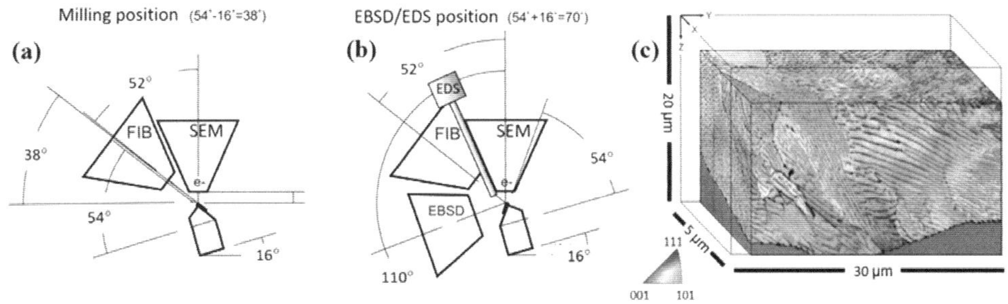

Figure 2. Automation of FIB-SEM EBSD and EDS. a) Schematics of FIB milling geometries. b) Schematics of EBSD/EDS data collection geometries. c) 3D EBSD volume is showing microstructure of perlite block. The colour is composed of image quality (grey value) and a colour code for the crystal direction parallel to the x-axis of the sample, after [20].

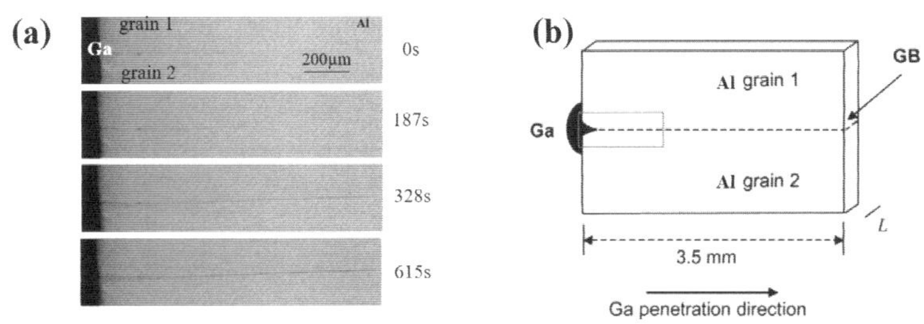

Figure 3. a) Synchrotron radiation X-ray microradiographs show continuous penetration and thickening of Ga along an Al grain boundary. b) Schematic drawing of the Al bicrystal in contact with the Ga droplet. The bicrystal has a grain boundary length of 3.5 mm and a width L of 0.8 mm [21].

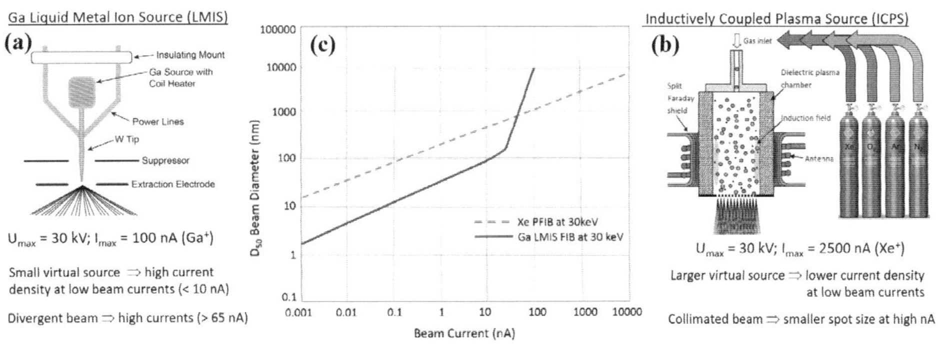

Figure 4. a) Schematics of liquid metal ion source. b) Schematics of inductively coupled plasma source. c) Comparison of the beam diameter versus the ion beam current for LIMS and ICPS based FIB columns [27].

In situ gas injection systems (GIS) have added site-specific matter deposition, micrometre-size complex structure fabrication [28, 29], and enhanced milling/etching functionalities [17]. While nano-manipulators together with GIS have enabled micro-scale sample preparation, manipulation and transfer, e.g., TEM lamella [30] and atom probe tomography (APT) [31], nano pillars extraction, preparation and lift-out [17].

FIB-SEM equipped with cryogenic stage kits are suitable for milling and imaging of untreated biological and beam sensitive specimens [32]. In Life Sciences, cryo FIB-SEMs are preparing samples for cryo electron correlative microscopy and tomography workflows [33].

In the early 2010's, the first experimental TriBeam systems, based on FIB-SEM platforms (Figure 5b), femtosecond laser (chirp pulse amplified (CPA) titanium sapphire (Ti:sapphire) femtosecond lasers with approximately 150 fs pulse width and 1 kHz repetition rate), and laser optics benches were tested independently at University of California Santa Barbara (UCSB), and FEI Company Hillsboro Oregon [34]. These systems had SEM, FIB, and femtosecond laser beams focussed on a coincidence point in the main vacuum chamber. From circa 2017, the FIB SEM with a separate vacuum chamber and femtosecond laser unit (Figure 5c) is also available commercially [35]. Although the applications envelope of fully functioning FIB-SEM is expanded to sub mm-scale, and to new capabilities, e.g., laser induced material deposition [34], 3D large volume material characterisation, etc., the aforementioned drawbacks of FIB are still remaining [34]. Particularly, the small material volume that can be accessed with high precision FIB milling is the apparent bottleneck in many advanced scientific and industrial applications.

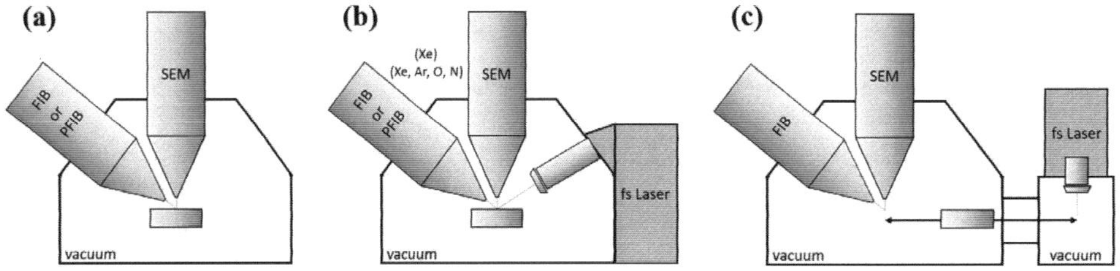

Figure 5. a) Schematics of FIB-SEM or PFIB-SEM. b) Schematics of integrated Laser FIB SEM or Laser PFIB-SEM. c) Schematics of FIB-SEM with an external laser unit and a separate vacuum chamber.

In order to solve the small material volume removal constraints, in 2012 the first Xe^+ plasma FIB SEM emerged commercially [36]. Typically, PFIB-columns use electron cyclotron resonance (ECR) [37] or inductively coupled plasma (ICP) source [38] (Figure 4b), and xenon noble gas to generate a Xe^+-ion beam. Use of ECR or ICP with heavy and inert Xe^+-ions suppress the two drawbacks of LMIS FIB-based instruments. ECR and ICP source produces broad and parallel beam, shaped, and focussed by the PFIB optics to a minimum spot sizes in dozens of nm range (pA beam currents). Although the imaging resolution of PFIB is worse than of FIB (see Figure 4c) or SHIM, the PFIB beam profile and spot size can be precisely shaped and controlled, even at very high beam currents [25] (Xe^+: 2,500 nA) and current densities. As a result, it greatly increases the maximum material removal rate (e.g., 675 $\mu m^3/s$ for Si [26], 38× faster than for FIB), and the maximum volume of material (for most of the materials ~ 200 × 200 × 200 μm^3, e.g., Figure 6) that can be removed with reasonable amount of time. For example, it takes ~ 25 minutes to remove a Si-cube measuring $(100\ \mu m)^3$ using the maximum material removal rate.

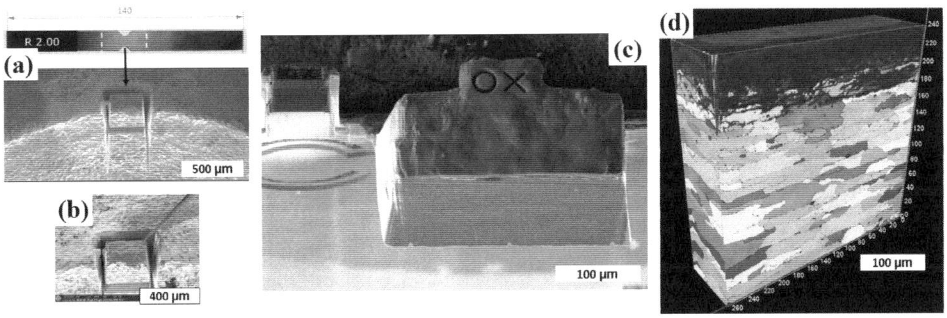

Figure 6. Large volume 3D EBSD collected using Plasma FIB-SEM [49]. a) Shot-peened double notch fatigue test 7075 sample with the material chunk ready for lift-out. b) The lift-out using EasyLift nano-manipulator. c) the material chunk attached to a Cu half-moon grid. d) 3D EBSD results (IPF-Z image stack. One of the volumetrically largest Xe^+ PFIB 3D EBSD datasets gathered to date.

Circa the year 2019, Xe^+-plasma FIB-SEM application envelope have been expanded by adding correlative microscopy (CM) holder kits/workflows [39], negative tilts for the stage, and installing three additional primary ion species for the PFIB column [40], namely argon, oxygen, and nitrogen (see Figure 4b). The CM holder kit solution allows multi-scale, multi-modal study, and precise tracking of volume of interest from X-ray microCT to DualBeams. The negative tilts allow for mimic the broad ion beam (BIB) polishing action at glancing angles while spinning the irradiated surface. This led to a new technique called PFIB spin milling, where sub-mm surface areas are sequentially removed with nm-precision and recorded with SEM qualitative, and quantitative imaging modes [41]. Two noble (Xe and Ar) and two reactive (O and N) gasses expand the Materials Science and Life Science applications range for the plasma FIB-SEM microscope [42, 43]. In the subsequent year, the low voltage PFIB performance have been improved, allowing precise Xe^+- and Ar^+- beam polishing of TEM samples at 1 keV [44] and 500 eV [45, 46].

In the year 2019 the first commercial system has emerged that expands the TriBeam arrangement concept (Figure 5b). The platform is equipped with monochromatic SEM and PFIB 2.0 columns, compact femtosecond laser (carbide ultra-short pulse (280 fs) laser tuned to green (515 nm) and near-infrared (1,030 nm) pulses), beam shaping and scanning optics [40], and negative tilts stage. This system can prepare large lamellas for TEM/TKD (Figure 7a), and mm range cross-sections and serial sections, and when equipped with fast analytical detectors it probes a mm^3-large 3D EBSD information [47, 48] (Figure 7b).

In 2021, two additional TriBeam configurations were introduced [50] to expand the application space of these systems further. The first is a PFIB-SEM system that can use multiple primary ion species, while the second is a Ga^+ FIB SEM system. Both systems are based on the Helios 5 platform and are equipped with the same compact femtosecond laser unit as the previous configurations (similar to Figure 5b).

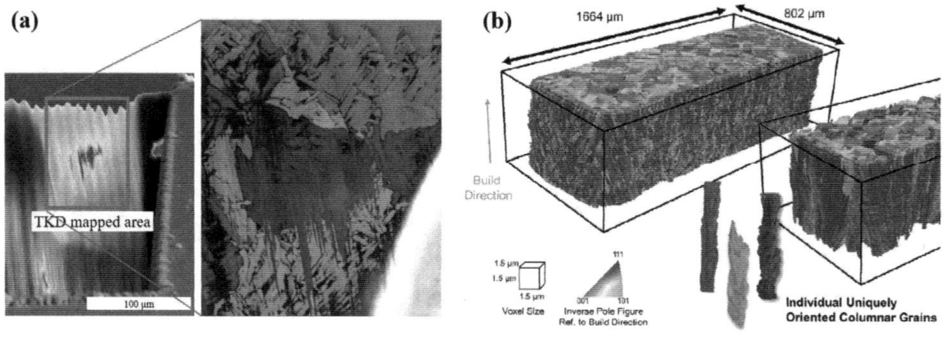

Figure 7. Helios 5 laser PFIB prepared samples. a) TKD lamella of tensile tested Ti64 alloys; Image courtesy: Dr Ali Gholinia, The University of Manchester. b) 3D reconstructions of an additively manufactured CoNi alloy sample. One of the volumetrically largest TriBeam datasets gathered to date [47].

2. Focussed beams and EBSD method

There are plenty of detectable and useful signals excited by a focussed electron beam in a solid matter (Figure 8a). High energy stationary electron beam is backscattered or transmitted (electrons are transmitted through thin film or lamella, typically < 100 nm thin) from diffracting crystallographic planes of a solid volume about 20 nm deep in the crystal material. These diffracted or transmitted electrons form on a detector (phosphor screen and CCD/CMOS camera, or directly recorded with a semiconductor direct electron detector) a regular pattern of curved bright lines, namely a backscattered Kikuchi diffraction (BKD) pattern, Figure 8c (the traces of Kossel cones on the detector plane, Figure 8b), or transmission Kikuchi diffraction (TKD).

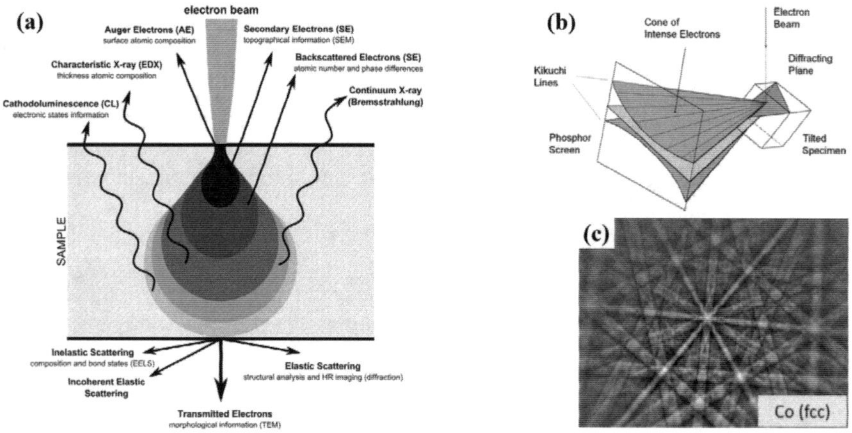

Figure 8. Interaction of focussed electron beam with a solid matter. a) Range of signals excited by an electron beam [51]. b) Schematic of the diffracting cones with respect to the reflecting plane, the specimen, and the phosphor screen [52]. c) Example backscattered Kikuchi diffraction pattern from Co grain [53].

The EBSD technique [54] (Figures 2b and 8b and 8c) and TKD [55-57] (Figures 9a and 9b) are standard methods to obtain spatially resolved crystallographic information via the automated analysis of Kikuchi diffraction patterns in the SEM. The modern EBSD method originated in 1993 when Dingley's research group published the first work on automated BKD analysis [58]. From the mid-1990's to the mid-2010's we can observe a steady increase in the number of publications that use phosphor/CCD-based EBSD technique (in practical applications of CCD-based EBSD data is collected and analysed with a speed < 1,000 fps) (Figure 10). The introduction of commercial phosphor/CMOS-based EBSD camera systems (in practical applications of CMOS based EBSD data collected and analysed with a speed < 2,000 fps) resulted in significant increase in the number of publications from circa 2017 (Figure 10). In 2013, the EBSD optical path for recording Kikuchi diffraction patterns is eliminated by experimental direct electron detection (DeD) systems [59]. DeD-based detector is a compact device, thus other EBSD geometries are possible, for example tilt-free EBSD [60], (Figs. 9c to 9e). The latest development in DeD-technology is electron energy-filtering detectors [61], and high-speed (~ 6,000 fps) BKD collection with the arbitrary kernel row addressing (AKRA) mode [62]. According to Scopus database [63], the first EBSD research work that uses FIB-SEM was published in 2002 [64] (Figure 11), laser FIB-SEM in 2012 [65], PFIB-SEM in 2016 [25], and laser PFIB-SEM in 2021 [47]. Further insight to the EBSD and TKD method, various experimental setups, electron detection technology, diffraction pattern simulations, pattern matching approaches, and the data post-processing is beyond the scope of this work. The author recommends for further reading books [52, 54], publications [52, 55-57, 59-62, 66-72], and the work cited there.

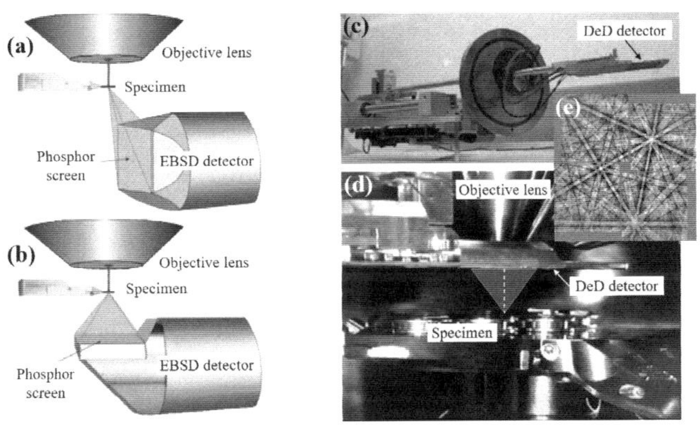

Figure 9. EBSD/TKD geometries. a) TKD using the standard EBSD camera, and b) the dedicated TKD camera [71]. An overview of the tilt-free DeD EBSD detector [60] with c) an image of the complete hardware design with electronics external to the SEM chamber, d) operational image of tilt-free geometry detector within the SEM chamber, and e) an example Ni EBSP collected in the tilt-free geometry at 20 kV with a 2 second exposure time.

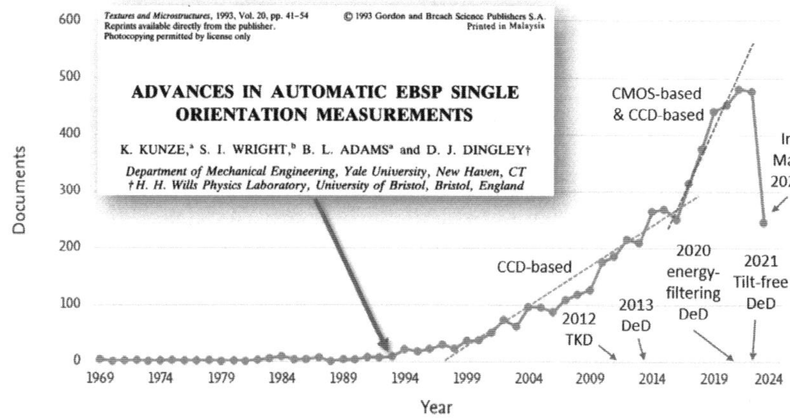

Figure 10. The graph shows number of documents published yearly on the EBSD related research, in total 5413 documents. Source: Scopus, May 2023. Search keywords: EBSD, electron backscattered diffraction, electron backscattering diffraction, backscattered Kikuchi diffraction, transmitted Kikuchi diffraction, TKD.

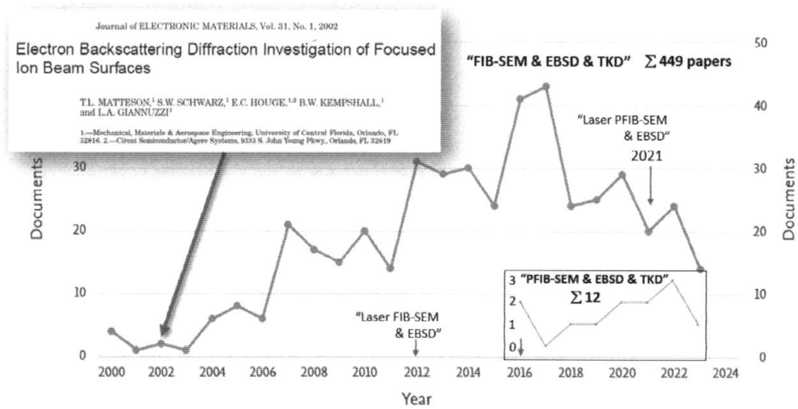

Figure 11. The graph shows number of research documents published yearly on the EBSD (TKD) integrated with FIB-SEM, PFIB-SEM, laser FIB-SEM, and laser PFIB-SEM. Source: Scopus, May 2023.

3. Surface preparation methods for EBSD

As mentioned in Section 2, Kikuchi diffraction patterns are formed by diffracted electrons travelling from the superficial layers of specimen's surface. Thus, the sample preparation methods, that retains an intact and planar at nanometre level crystalline microstructure, are critical to achieve unbiased measurements. Frequently, specimen surfaces are covered with contaminants, oxides, surface reaction products, and crystallographic scale damage induced by plastic deformations, amorphisations, heat or implantation damage. Ceramics, semiconductors, metals, and their composites, e.g., cermets, respond differently to the common sample preparation methods. Therefore, materials and specimens need to be prepared on an individual basis, and it may involve a "trial-and-error" approach.

IOP Conf. Series: Materials Science and Engineering **1324** (2025) 012013 doi:10.1088/1757-899X/1324/1/012013

A number of sample preparation methods for EBSD/TKD is available, and have been used [17, 30, 31, 41, 64, 65, 73-79] (these references point to one of the first papers published on the subject matter): a) metallographic grinding and polishing, b) electropolishing, c) chemical polishing, d) cleaving and fracturing, e) broad ion beam planar and cross-sectioning polishing, f) focussed ion beam planar (FIB or plasma FIB planar polishing is known as FIB/PFIB spin milling [41), cross-sectioning polishing, g) and femtosecond laser polishing. This paper gives a brief insight on methods f) and g).

3.1. Ga^+ FIB and plasma FIB

A high-energy beam of particles interacts with a solid, causing initiation of several processes in the volume of interaction [17], as depicted in Figure 12a. When considering a focussed beam preparation of the sample surface, it is crucial to maximise*: i*) the sputtering of substrate's atoms (the maximums in Figure 12b), and *ii*) the backsputtering of primary ions.

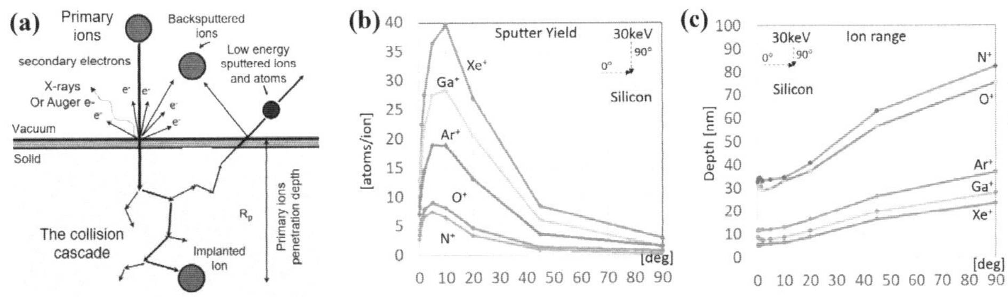

Figure 12. a) Schematic diagram of the sputtering process and ion-solid interactions (based on [80]). SRIM simulation of b) sputter yield, and c) ion penetration range of several primary ions species interacting with a silicon at various incidence angles [39].

At the same time, *iii*) the collisions induce secondary processes that need to be suppressed to a negligible level. These are: primary ions implantation (the minimums in Figure 12b), chemical reactions, recoils of constituent atoms, defects formation, thermal and radiation-induced diffusion, mixing of constituent elements, phase transformation, amorphisation, and re crystallisation. Further, it is important to suppress *iv*) the surface morphology changes by the ion implantation and sputtering, namely, curtains, craters, facets, grooves, ridges, pyramids formation, blistering, and exfoliations. In general, processes *i*) - *iv*) are competing phenomena. Thus, to achieve EBSD/TKD suitable surface finish the following priorities for sample preparation need to be enforced: 1) pristine surface finish and roughness at nm-range, 2) intact crystal lattices, and 3) maximum sputter yield of substrate atoms. As a result, low glancing angles (0° - 8°.) of FIB, PFIB milling and polishing, and a range of ion energies (30 keV - 500 eV) are used (Figure 13a). PFIB may need additionally application of the rocking mill approach to mitigate surface curtains [25]. Automated 3D EBSD runs are typically performed at 30 keV and the polishing currents (FIB: 1 nA, PFIB: 60 nA). EBSD/TKD sample preparation is starting from the high ion energy (30 keV) and beam milling/polishing currents (FIB: 10 … 1 nA, PFIB: 200 … 60 nA), and finishes at very low ion energies (down to 500 eV, for FIB and PFIB) and beam currents (≥ 50 pA). SRIM [81] calculated ions ranges in silicon for Ga+ FIB and PFIB (Ar^+, Xe^+) are similar for energies < 2 keV, but PFIB gives the benefit of gallium-free specimen preparation (Figure 13b). Ar^+ PFIB prepared samples directly compete with Ar^+ broad ion beam polished specimens [46]. While recently, Ne^+ HIM have been used to prepare an EBSD quality Cu specimen [82].

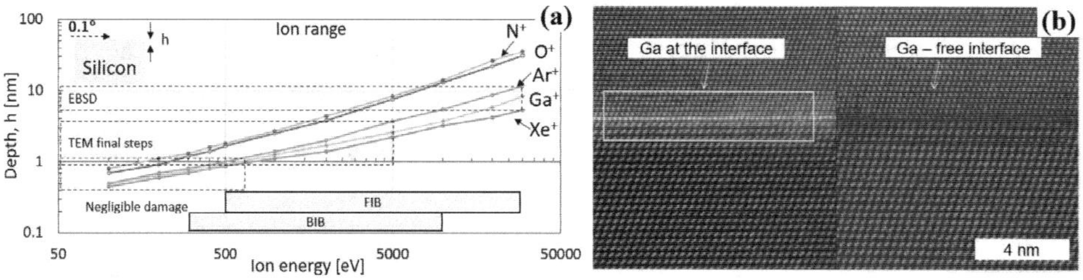

Figure 13. a) SRIM simulation of ions implantation ranges for energies 100 eV to 30 keV; 0.1 glancing angle of ions. b) Titan 300, STEM HAADF Z-contrast; Al/Si interface; Ga FIB versus Xe PFIB. Image courtesy: Brandon Van Leer and Chengge Jiao, Thermo Fisher Scientific [46].

3.2. Ultra-short pulse lasers (USPL)

In simplified terms, laser electromagnetic radiations interact only with electrons of material atoms, since nuclei are too heavy [83]. The processes vary for metals, semiconductors, and dielectrics. Figure 14 shows the timescale of electron and lattice processes in laser-irradiated solids. Femtosecond laser pulse transfers energy during carrier excitation and thermalisation, subsequent processes follow without external energy. While being close to the ablation threshold, the other thermal and structural effects are suppressed to a negligible level, allowing high quality EBSD mapping [34, 47, 65, 84, 85], see Figures 15b and 15c (adopted from [84]). Very high fluences transfer so much energy causing pronounced thermal and structural effects [34, 86, 87], e.g., dislocations injection, phase transformation, amorphisation and re-crystallisation.

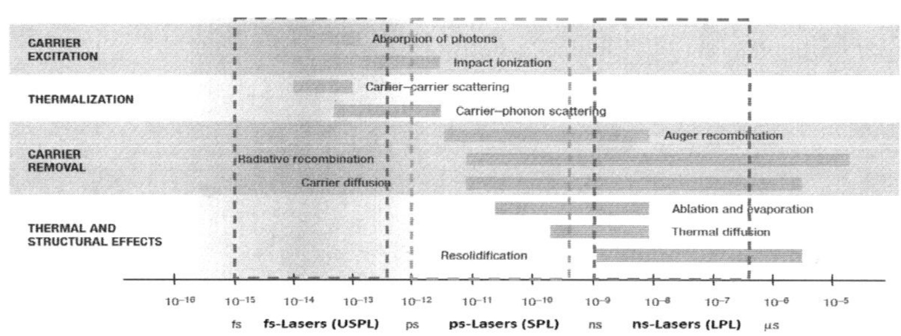

Figure 14. Timescale of electron and lattice processes in laser-irradiated solids [83].

One of inherent features of femtosecond laser ablation of a solid in the low fluence regime is formation of the laser induced periodic surface structure (LIPSS) [84, 85], see regions A in Figure 15b. LIPSS varies significantly with the pulse width of the laser beam, and laser processing conditions (the laser polarisation, repetition rate, laser fluence, laser scanning patterns) [47, 48]. These nano-sized structures retain the original crystalline lattice with negligible lattice damage (see Figure 7a) have the periodicity in the range of half of the laser wavelength, and the roughness in the range of dozens of nanometres. LIPSS may obscure the diffracted electrons leading to non-indexed areas on the surface (Figure 16a). However, tedious selection of laser processing parameters can suppress LIPSS to the negligible level (Figure 16a) [88]. Materials surfaces prone to laser damage need to be post-laser processed. For example, the surface damage and the periodic surface structure can be removed by FIB (30 keV … 5 keV, < 30 nA, < 10° glancing angle [85], Figure16b) or PFIB (30 keV and lower, ≤ 60 nA,

< 10° glancing angle [26]) polishing. Nevertheless, some materials (e.g., GaN [85]) are more sensitive to FIB induced damage than damage generated by fs-laser ablation, leading to reduced quality of recorded Kikuchi patterns.

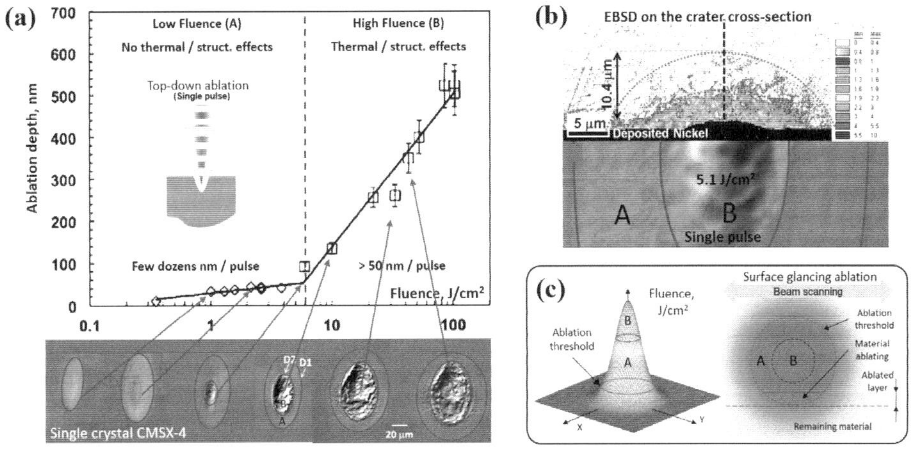

Figure 15. Femtosecond laser interaction with matter. a) The ablation depth versus the laser fluence for CMSX-4 Ni-based alloy single crystal [84]. b) EBSD grain reference orientation deviation (GROD) map (in degrees) [84]. c) Schematics of laser fluence spatial profile and ablation geometry at surface glancing angles (based on [85]).

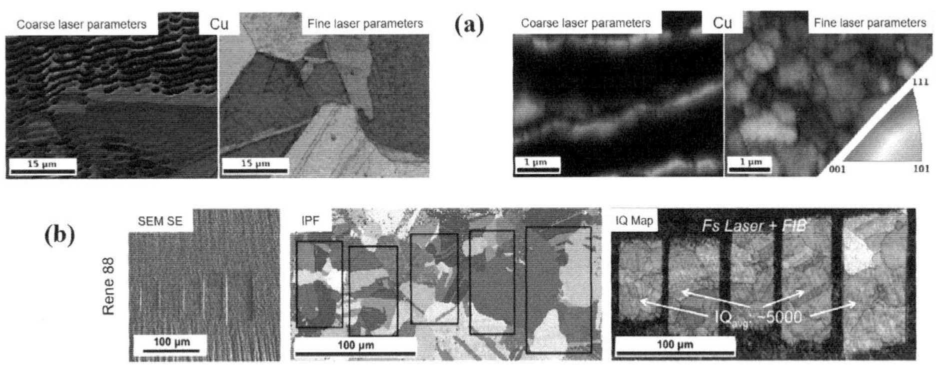

Figure 16. EBSD images (inverse pole figure with superimposed image quality). a) Coarse grain copper (left) and ultrafine grain cooper processed with a 300 fs, $\lambda = 512$ nm laser, the fluence 1.96 J/cm^2, the repetition rate 37.5 kHz for coarse, and 12.5 kHz for fine laser parameters [88]. b) Rene 88 machined with 150 fs, $\lambda = 780$ nm laser with 1 kHz repetition rate, and FIB polished at 30 keV, 20 nA [85].

4. Conclusion and further directions

This paper briefly reviews the last 30 years of the development of the focussed beams (electron, ion, and photon beams) and their application for the EBSD and TKD methods. In the beginning, these beams formed standalone research platforms. The development of FIB-SEM microscope opened up in situ, and site-specific investigation in the micro- and nano-scale, and in two- and three-dimensions. Improved beams performance, various electron detection channels and analytical detectors (EBSD, EDS, WDS, etc.) add new imaging modalities and quantitative analyses possibilities. Xenon plasma FIB expands the application space, and increases 3D EBSD material volume analysed with DualBeams. EBSD and EDS data mapping remain the slowest part of the 3D data collection workflow, although there are exciting improvements in detector speed with DeD EBSD. Multi-ion plasma beam and low keV beam performance further expands the application spaces of PFIB to a) a broader range of life sciences applications (with oxygen beam), and b) near damage-free surface preparation (with argon and xenon beams). The recent addition of femtosecond laser beam to the Helios 5 DualBeam microscope platforms enables the collection of high-quality 3D EBSD datasets in the mm^3-volume range at sub-micrometre SEM imaging resolution. Athermal femtosecond laser serial sectioning of soft polymers and biological materials, as well as integrated circuit devices, demonstrates good surface quality and finish. PFIB-SEM and laser PFIB-SEM becoming an essential part in the multi-modal CM workflows, being the essential link between the macro- and nano-scale. The envisaged advancement of CM workflows, on-the-fly data reconstruction, smart sampling, active prior knowledge analysis and instrument feedback, use of machine learning, as well as additional imaging modalities, will expand the already broad application envelope of the focussed beams.

References

[1] Ruska E 1986 *Appl. Optics* **25** 820
[2] Hawkes P W 1985 *The beginnings of electron microscopy.* (Hawkes P W; Ed.) [London: Academic Press]
[3] Von Ardenne M 1939 *Improvements in electron microscopes.* Patent GB511204A [UK]
[4] McMullan D 1995 *Scanning* **17** 175-185
[5] Krohn V E and Ringo G R 1975 *Appl. Phys. Lett.* **27** 479-481
[6] Orloff J 1996 *J. Vacuum Sci. Technol. B* **14** 3759
[7] Jun-ichi Nishizawa M J A 2009 *Proc. Jpn Acad. Ser. B: Phys. Biol. Sci.* **85** 454-465
[8] Schawlow A L and Townes C H 1958 *Phys. Rev.* **112** 1940
[9] Sooy W 1965 *Appl. Phys. Lett.* **2** 36-58
[10] Goldstein J, *et al.* 2017 *Scanning electron microscopy and X-ray microanalysis.* [Berlin-Heidelberg: Springer-Verlag]
[11] Hawkes P W and Spence J C H (Eds.) 2019 *Springer handbook of microscopy.* [Berlin-Heidelberg: Springer]
[12] Zeiss C 2008 *Carl Zeiss SMT ships world's first ORION helium ion microscope to U.S. National Institute of Standards and Technology.* [cited 2008 July 17]
[13] GlobeNewswire 2011 *FEI announces new Vion PFIB system for advanced IC packaging.* [cited 2011 June 13]. Faster material removal opens new markets for failure analysis in advanced integrated circuit packaging applications.
[14] Knuffman B, Steele A V and McClelland J J 2013 *J. Appl. Phys.* **114** 044303
[15] Pilz W, *et al.* 2019 *J. Vacuum Sci. Technol. B* **37** 021802
[16] Bassim N, Scott K and Giannuzzi L A 2014 *MRS Bull.* **39** 317-325
[17] Yao N 2010 *Focussed ion beam systems, basics and applications.* [Cambridge, MA: Cambridge University Press]
[18] Feng Q, *et al.* 2005 *Scripta Materialia* **53** 511-516

[19] Williams J. 2014 *The dual-beam - Its birth, progress, and future.* in: *Microscopy Today*, July 2014
[20] Zaefferer S, Wright S I and Raabe D 2008 *Metallurgical Mater. Trans. A: Phys. Metall. Mater. Sci.* **39** 374-389
[21] Ludwig W, Pereiro-López E and Bellet D 2005 *Acta Materialia* **53** 151-162
[22] Lu N, *et al.* 2021 *Acta Materialia* **217** 117145
[23] Hugo R C and Hoagland R G 1998 *Scripta Materialia* **38** 523-529
[24] Lejcek P, Sob M and Paidar V 2017 *Progress Mater. Sci.* **87** 83-139
[25] Burnett T L, *et al.* 2016 *Ultramicroscopy* **161** 119-129
[26] Winiarski B and Geurts R 2020 *Microsc. Microanal.* **26** (Suppl 2) S4-S8
[27] Smith N S, *et al.* 2006 *J. Vacuum Sci. Technol. B: Microelectr. Nanometer Struct.* **24** 2902-2906
[28] Jenke M G, *et al.* 2011 *Nano Letters* **11** 4213-4217
[29] Prestigiacomo M, *et al.* 2004 *Microelectr. Eng.* **76** 175-181
[30] Jacob P, *et al.* 1999 *New FIB-supported approaches for EELS-capable TEM-lamella preparation.* in: Proc. Int. Symp. Physical and Failure Analysis of Integrated Circuits (IPFA). 69-72
[31] Thompson G B, *et al.* 2003 in: *Proc. Materials Research Society Symp.* **788** 553-558
[32] Mulders H 2003 *GIT Imaging Microsc.* **2** 8-10
[33] Schaffer M, *et al.* 2015 *Bio Protoc.* **5** 1575
[34] Echlin M P, *et al.* 2015 *Mater. Characterization* **100** 1-12
[35] Pfeifenberger M J, *et al.* 2017 *Materials & Design* **121** 109-118
[36] Jiruše J, *et al.* 2012 *Microsc. Microanal.* **18** 652-653
[37] Boukari F, *et al.* 1994 *Rev. Scientif. Instrum.* **65** 1097-1099
[38] Nabhiraj P Y, *et al.* 2010 *Nucl. Instrum. Meth. Phys. Res. A: Accel.rators, Spectrom., Detectors Assoc. Equipm.* **621** 57-61
[39] Winiarski B, *et al.* 2019, *Advances in multi-beam and multi-ion FIB-SEM for 3D correlative microscopy.* in: Microscopy and Microanalysis 2019 Meeting (Portland, OR)
[40] Winiarski B, *et al.* 2019 *Microsc. Microanal.* **25** (Suppl 2) 870-871
[41] Winiarski B, Rue c and Withers P J 2019 *Microsc. Microanal.* **25** 350-351
[42] Eder K, *et al.* 2021 *Ultramicroscopy* **228** 113334
[43] Wang J, *et al.* 2021 Sci. Reports **11** 13162
[44] Van Leer B, *et al.* 2020 *Microsc. Microanal.* **26** (Suppl 2) 898
[45] Jiao C 2021 *Towards "damage-free" TEM specimen preparation by focused ion beam without gallium.* in: MRS OnDemand 2021 [Material Research Society]
[46] Jiao C, *et al.* 2021 *Low energy 500 eV focused argon ion beam provided by multi-ions species plasma FIB for Material Science sample preparations.* in: Microscopy and Microanalysis Meeting 2021. [Cambridge, MA: Cambridge University Press]
[47] Echlin M P, *et al.* 2021 *J. Materials* **73** 4258-4269
[48] Echlin M P, *et al.* 2020 *Curr. Opin. Solid State Mater. Sci.* **24** 100817
[49] Winiarski B, Withers P J and Burnett T L 2016 *Xe^+ plasma FIB milling and lift-out approach for site-specific preparation of large volume blocks for 3D-EBSD.* in: Microscopy and Microanalysis Meeting 2016. (Columbus, OH) [Cambridge, MA: Cambridge University Press]
[50] Thermo Fisher Scientific. *Commercial Announcement: Helios 5 Laser DualBeam and Helios 5 Laser Hydra DualBeam.* in: Microscopy and Microanalysis Meeting 2021. [Cambridge, MA: Cambridge University Press]
[51] Claudionico 2013 *Illustration of the phenomena that occur from the interaction of highly energetic electrons with matter, also depicting the pear shape interaction volume which is typically observed in this type of interactions.* Available from: https://commons.wikimedia.org/wiki/File:Electron_Interaction_with_Matter.svg

[52] Schwarzer R A, *et al.* 2009 *Present state of electron backscatter diffraction and prospective developments.* in: Electron backscattered diffraction in Materials Science. (Schwartz A J, et al.; Eds.) [New York, NY: Springer]
[53] Winiarski B, *et al.* 2021 *Ultramicroscopy* **226** 113315
[54] Schwartz A J, Kumar M and Adams D P 2000 *Electron backscatter diffraction in Materials Science.* [New York, NY: Kluwer Academic / Plenum Publications]
[55] Keller R R and Geiss R H 2012 *J. Microscopy* **245** 245-251
[56] Trimby P W and Cairney J M 2014 *Adv. Mater. Processes* **172** 13-15
[57] Liu J, *et al.* 2019 *Ultramicroscopy* **205** 5-12
[58] Kunze K, *et al.* 1993 *Textures Microstructures* **20** 41-54
[59] Wilkinson A J, *et al.* 2013 *Phys. Rev. Lett.* **111** 065506
[60] Marshall A L, *et al.* 2021 *Ultramicroscopy* **226** 113294
[61] Bammes B and Bilhorn R 2020 *Microsc. Microanal.* **26** 1186
[62] Wang F, *et al.* 2021 *Ultramicroscopy* **220** 113160
[63] SCOPUS 2022 Keywords: FIB-SEM or FIB and EBSD. (cited 2022 28-Feb-2022). Available from: https://www.scopus.com
[64] Matteson T L, *et al.* 2002 *J. Electr. Mater.* **31** 33-39
[65] Echlin M P, *et al.* 2012 *Rev. Scientif. Instrum.* **83** 023701
[66] Winkelmann A, *et al.* 2021 *J. Microscopy* **284** 157-184
[67] Singh S, Ram F and de Graef M 2017 *J. Appl. Crystallogr.* **50** 1664-1676
[68] Foden A, *et al.* 2019 *Ultramicroscopy* **207** 112845
[69] Trimby P W 2012 *Ultramicroscopy* **120** 16-24
[70] Mingard K P, *et al.* 2018 *Ultramicroscopy* **184** 242-251
[71] Schwarzer R 2021 *Transmission Kikuchi diffraction TKD. EBSD and BKD.* (cited 2022). Available from: http://www.ebsd.info/tkd.htm
[72] Guyon J, *et al.* 2016 *Ultramicroscopy* **161** 161-167
[73] Echlin P 2009 *Handbook of sample preparation for scanning electron microscopy and X ray microanalysis.* [Cambridge, MA: Springer]
[74] Katrakova D and Mücklich F 2001 *Prakt. Metallographie/Pract. Metallography* **38** 547-565
[75] Ajantiwalay T, *et al.* 2019 *Micron* **118** 1-8
[76] Nowell M M, Witt R A and True B W 2055 *EBSD sample preparation: Techniques, tips, and tricks. Microscopy Toda*y **13**(4) 44-49
[77] Wynick G L and Boehlert C J 2005 *Mater. Characterization* **55** 190-202
[78] Erdman N, Campbell R and Asahina S 2066 *Precise SEM cross-section polishing via argon beam milling. Microscopy Today* **14** 22-25
[79] Michael J R and Kotula P G 2008 *Microsc. Microanal.* **14** (Suppl 2) 976-977
[80] Giannuzzi L A and Stevie F A (Eds.) 2005 *Introduction to focused ion beam.* [New York, NY: Springer]
[81] Ziegler J, Biersack J and Littmark U 1985 *The stopping range of ions in matter.* [New York, NY: Pergamon Press]
[82] Wolff A 2021 *Microsc. Microanal.* **27** (Suppl 1) 1016
[83] Sundaram S K and Mazur E 2002 *Nature Mater.* **1** 217-224
[84] Kumar A and Pollock T M 2011 *J. Appl. Phys.* **110** 083114
[85] Echlin M P, *et al.* 2017 *Acta Materialia* **124** 37-46
[86] Gorelik T, *et al.* 2003 *Appl. Phys. A: Mater. Sci. Processing* **76** 309-311
[87] Sedao X, *et al.* 2016 *ACS Nano* **10** 6995-7007
[88] Jelinek A, *et al.* 2021 *J. Materials* **732** 4248-4257

Electron probe microanalysis of UO_2 for stoichiometry determination

K E Wright and C Adkins

Idaho National Laboratory, Advanced Characterization and PIE, P.O. Box 1625
MS 6000, US-83415 Idaho Falls, Idaho, U.S.A.

E-mail: karen.wright@inl.gov

Abstract. Oxygen-metal ratios are important to the study of irradiated nuclear oxide fuel. Critical performance metrics such as thermal conductivity, constituent redistribution, and melting temperature depend on the degree of departure from $UO_{2.00}$ stoichiometry and in which direction (hyper- or hypo-stoichiometric) the departure has occurred. This study examined three sintered pellets with compositions of $UO_{2.00}$, $UO_{2.02}$, $UO_{2.21}$ and one un-sintered pellet with a composition of $UO_{2.16}$. When analysed via electron probe microanalysis, Al-coated pellet accuracy ranged from -1.0 % to 19.4 % while Ir-coated pellet accuracy ranged from -0.5 % to 4.6 %. For both analysis sets the un-sintered pellet exhibited the largest accuracy error. While some pellet heterogeneity contributed to lower-than-expected precision, this method appears suitable for characterising oxygen-metal ratios in irradiated uranium oxide fuels.

1. Introduction

Nuclear fuel oxide-metal (OM) ratios are important for predicting oxide fuel behaviour. Oxide fuels can be fabricated stoichiometric (i.e., $UO_{2.00}$) or non-stoichiometric (e.g., $UO_{1.98}$) depending on the desired behaviour and reactor type. However, due to fission product, oxygen, and actinide element redistribution, stoichiometry along the pin radius evolves during irradiation in ways that affect performance [1]. For example, thermal conductivity is at its maximum for stoichiometric UO_2 but decreases with hypo- and hyper-stoichiometry [1, 2]. In fast reactor mixed-oxide fuel (MOX), the fuel restructuring rate was accelerated in stoichiometric fuel compared to that of hypo-stoichiometric fuel, even though the latter fuel experienced a higher temperature [3]. Stoichiometric UO_2 melts congruently, and at higher temperatures than non-stoichiometric UO_2 [2], impacting fuel behaviour during off-normal events.

Numerous methods have been suggested or tested to determine the OM ratio in nuclear fuel (see [4] for a review of approximately 35 such methods) with thermogravimetric analysis and X-ray diffraction employed commonly. The disadvantage of these latter two methods is that they lack sufficient spatial resolution to determine the OM ratio on a micrometre spatial scale. In addition, these methods cannot provide elemental fuel and fission product analysis along the radius, nor can they be used to study microstructural changes occurring during irradiation.

In the early 1970s scientists began using electron probe microanalysis (EPMA) to study fuel element and fission product redistribution in irradiated nuclear fuel at a micrometre spatial scale. Shielded EPMAs were developed to protect electronics from radiation-induced degradation while simultaneously improving trace element detection limits. Since they are equipped with electron imaging detectors, shielded EPMAs are also used to study microstructural changes in irradiated fuel. Because of EPMA's utility in irradiated fuel analysis, it is potentially valuable for measuring OM ratios and furthering our understanding of irradiated fuel behaviour, particularly as new types of nuclear fuels are introduced and tested.

This article represents the first step to developing a suitable EPMA OM measurement method— that of determining the accuracy resulting from measuring UO_2 of varying stoichiometries.

2. Methods

Powdered $UO_{2.00}$, $UO_{2.02}$, and $UO_{2.21}$ were fabricated into pellets, pressed (with no binders) at 50 MPa, and sintered at 1,050 °C for 4 hours. The sintering chamber atmospheric gas was controlled to maintain the required pO_2 to preserve the fabricated stoichiometry. $UO_{2.16}$ was fabricated in the same manner but was not sintered. The pellets were subsampled and measured using thermogravimetric analysis to determine bulk pellet stoichiometry [5]. A subsection of each pellet was then mounted in epoxy and polished to a surface finish of < 1 μm using diamond paste. To render their surfaces conductive, pellets and a $UO_{2.00}$ standard were coated simultaneously with 15 nm of aluminium.

At least thirty defocussed spots were measured on each pellet with a CAMECA SX100R EPMA and were matrix-corrected with the Pouchou and Pichoir [6] phi-rho-z correction. The mass absorption coefficient of O-kα in U used was that of Pöml and Llovet [7]. Because aluminium is more easily oxidised than iridium, the pellets were polished to remove the Al-coating, and the experiment was repeated using a 10 nm-thick Ir conductive coating on the same standard and pellets. While a 10 nm-thick Ir coating is excessively thick for many applications, that thickness is necessary for irradiated nuclear fuel due to its extreme porosity. Measurement conditions are shown in Table 1.

Table 1. $UO_{2\pm x}$ EPMA measurement conditions.

X-ray line	Peak count time (s)	Background count time (s)	keV	nA	Diffraction crystal
U-Mα	60	30	20	40	Quartz
O-Kα	60	30	20	40	PC1

3. Results

Each pellet's stoichiometry was measured by bulk thermogravimetry [5]. However, the pellets were less homogeneous on a micrometre spatial scale than was considered optimal for this study. Because of this, each pellet's EPMA-measured OM points were averaged and points lying outside one standard deviation of the average were omitted from the data set. Results for each pellet are shown in Table 2 for Al-coated pellets and Table 3 for Ir-coated pellets.

Table 2. Shows average ($n \geq 30$) UO_x values and accuracy deviation for Al-coated pellets.

Predicted UO_2 measurement	Measured UO_2 (Al-coated)	% deviation from predicted value
2.00	2.02	1.0
2.02	2.00	-1.0
2.16	2.58	19.4
2.21	2.26	2.3

Table 3. Shows average ($n \geq 30$) UO_x values and accuracy deviation for Ir-coated pellets.

Predicted UO_2 measurement	Measured UO_2 (Ir-coated)	% deviation from predicted value
2.00	2.00	0.0
2.02	2.01	-0.50
2.16	2.26	4.6
2.21	2.23	0.9

4. Discussion

When examining Tables 2 and 3, it is apparent that the measured average OM ratio for $UO_{2.00}$, $UO_{2.02}$ and $UO_{2.21}$ pellets is quite close to each pellet's respective predicted value. Al-coated pellet measurement accuracy ranged from -1.0 % to 2.3 % for those three pellets, while Ir-coated measurement accuracy ranged from -0.5 % to 0.9 % for the same three pellets. However, for both Al- and Ir-coated $UO_{2.16}$ pellets, the departure from the predicted value was quite large. The Al-coated pellet measured 19.4 % higher than its predicted value, while the Ir-coated pellet measured 4.6 % higher than its predicted value. Since the $UO_{2.16}$ pellet had not been sintered, this data suggests that the greater surface area inherent with un-sintered material significantly affected pellet oxidation and/or oxygen adsorption onto the pellet surface. Moreover, this data shows that when measuring oxygen, it is helpful to use a conductive coating such as Ir, which is less readily oxidised.

As demonstrated by the average OM measurements shown in Tables 2 and 3, when the pellet was sintered, the average measured UO_x-value was reasonably accurate regardless of the coating used in this study; however, precision is important because it permits distinguishing measurements that are statistically within error from those that are statistically different. As mentioned in Section 3, these pellets were less homogeneous than would be ideal for this study. This presented the challenge that measurement differences could be due to uncertainties inherent to the measurement process, or they could be due to sample heterogeneity, or both. Figure 1 (a-b) shows 3-sigma (99 % confidence interval) error bars as a percentage of the average measurement value. This error encompasses both pellet inhomogeneity and measurement method error. When comparing the Al-coated pellet measurements with the Ir-coated pellet measurements, the relative size of the error bars is similar. For example, for both coatings, the smallest error is observed with $UO_{2.00}$ while the largest error is seen with $U_{2.16}$.

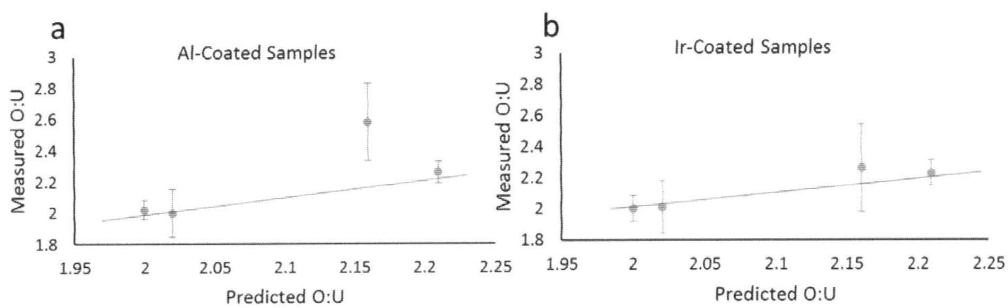

Figure 1. a) Shows measured O:U ratios compared with predicted values for stoichiometric and non-stoichiometric Al-coated UO_2. b) Shows measured O:U ratios compared with predicted values for stoichiometric and non-stoichiometric Ir-coated UO_2. The one-to-one correlation line is shown in red. Error bars represent the 99 % confidence interval.

Because the error bar magnitude includes both method error and pellet heterogeneity, it likely portrays lower precision than would typically occur with a perfectly homogeneous pellet. Despite this shortcoming, $UO_{2.00}$ and $UO_{2.02}$ were statistically different from $U_{2.21}$ for both data sets, though the former two were not statistically different from each other. With the Al-coated pellets, all the sintered pellets are statistically different from the un-sintered pellet mainly because the measured average value of the un-sintered pellet was 19.4 % higher than the predicted value, while the other three pellets' accuracy was considerably better.

As this method is intended for use with irradiated oxide fuels, accuracy is more achievable than precision. Irradiated fuels are inhomogeneous on a micrometre spatial scale and exhibit considerable porosity making high-precision measurements difficult to achieve. However, typically ~ 200 measurements are collected along a fuel's radius. This method serves to characterise accurately OM average values and trends in uranium oxide fuels when many point measurements are collected. Further research is required to determine whether this method is similarly accurate for MOX fuel.

5. Conclusions

- Ir-coating is less susceptible to oxidation and yields more accurate UO_x stoichiometry measurements than are achieved when using an Al-coating.
- In sintered pellets, measurements were accurate to within approximately 1 % (relative) for Ir-coated pellets and within approximately 2 % (relative) for Al-coated pellets.

Acknowledgements
This work was sponsored by the U.S. Department of Energy, Office of Nuclear Energy, under U.S. Department of Energy Idaho Operations Office Contract DE-AC07-05ID14517, as part of the National Scientific User Facility instrument scientist programme. The United States Government retains and the publisher, by accepting the article for publication, acknowledges that the United States Government retains a nonexclusive, paid-up, irrevocable, world-wide license to publish or reproduce the published form of this manuscript, or allow others to do so, for United States Government purposes.

References

[1] Matsumoto T, Arima T, Inagaki Y, Idemitsu K, Kato M and Uchida T 2013 *J. Nucl Mater.* **440** 580-585
[2] Olander D 1976 *Fundamental aspects of nuclear reactor fuel elements.* United States, https://doi.org/10.2172/7343826. https://www.osti.gov/servlets/purl/7343826
[3] Ikisawa Y, Maeda K, Kato M and Uno M 2017 *Nucl Tech* **199:1** 83-95
[4] Rubin J, Chidester K and Thompson M 2000 *O/M ratio measurements in pure and mixed oxide fuels – where are we now?* Report: LA-UR-00-5805. [Los Alamos, NM: Los Alamos National Laboratory]
[5] ASTM 2016 *Standard test method for determination of the oxygen to metal (O/M) ratio in sintered mixed oxide ((U, Pu)O₂) pellets by gravimetry.* ASTM C1817-16. [West Conshohocken, PA: ASTM]
[6] Pouchou J and Pichoir F 1991 in: *Electron probe quantification.* (Heinrich K F J and Newbury D E; Eds.). [New York,NY: Plenum Press] 31-76
[7] Pöml P and Llovet X 2020 *Micros Microanal.* **26** 194-203

AUTHOR INDEX

Institute of Physics
Dirac House, Temple Back
Bristol BS1 6BE UK

ISSN: 1757-8981
ISBN 979-8-3313-1417-0

CURRAN ASSOCIATES INC.
proceedings.com